D1357411

Samuel Beckett

New Interpretations of Beckett in the Twenty-first Century

As the leading literary figure to emerge from post–World War II Europe, Samuel Beckett's texts and his literary and intellectual legacy have yet to be fully appreciated by critics and scholars. The goal of *New Interpretations of Beckett in the Twenty-first Century* is to stimulate new approaches and develop fresh perspectives on Beckett, his texts, and his legacy. The series will provide a forum for original and interdisciplinary interpretations concerning any aspect of Beckett's work or his influence upon subsequent writers, artists, and thinkers.

Samuel Beckett: History, Memory, Archive
edited by Seán Kennedy and Katherine Weiss

Beckett's Masculinity
by Jennifer M. Jeffers

Samuel Beckett
History, Memory, Archive

Edited by

Seán Kennedy and Katherine Weiss

SAMUEL BECKETT
Copyright © Seán Kennedy and Katherine Weiss, 2009.

First published in 2009 by
PALGRAVE MACMILLAN®
in the United States—a division of St. Martin's Press LLC,
175 Fifth Avenue, New York, NY 10010.

Where this book is distributed in the UK, Europe and the rest of the world,
this is by Palgrave Macmillan, a division of Macmillan Publishers Limited,
registered in England, company number 785998, of Houndmills,
Basingstoke, Hampshire RG21 6XS.

Palgrave Macmillan is the global academic imprint of the above companies
and has companies and representatives throughout the world.

Palgrave® and Macmillan® are registered trademarks in the United States,
the United Kingdom, Europe and other countries.

ISBN: 978–0–230–61944–9

Library of Congress Cataloging-in-Publication Data

Samuel Beckett : history, memory, archive / edited by Seán Kennedy
and Katherine Weiss.
    p. cm.—(New interpretations of Beckett in the twenty-first century)
    ISBN 0–230–61944–4 (alk. paper)
    1. Beckett, Samuel, 1906–1989—Criticism and interpretation.
I. Kennedy, Seán, 1974– II. Weiss, Katherine.

PR6003.E282Z8177 2009
848'.91409—dc22   2009012212

A catalogue record of the book is available from the British Library.

Design by Newgen Imaging Systems (P) Ltd., Chennai, India.

First edition: November 2009

10 9 8 7 6 5 4 3 2 1

Printed in the United States of America.

Transferred to Digital Printing in 2010

Contents

Series Editor's Foreword

As the leading literary figure to emerge from post–World War II Europe, Samuel Beckett's texts and his literary and intellectual legacy have yet to be fully explored by critics and scholars. The purpose of "New Interpretations of Beckett in the Twenty-First Century" is to stimulate new approaches and fresh perspectives on Beckett's texts and legacy. The series will provide a forum for original and interdisciplinary interpretations concerning Beckett's work and/or his influence upon subsequent writers, artists, and thinkers. Much has been made of James Joyce's influence on Beckett (which is limited to the early years of his career), but there has yet to be a thorough analysis of Beckett's influence not only on writers (Vaclav Havel, Edna O'Brien, Harold Pinter, J. M. Coetzee, and James Kelman) but also on artists (Jasper Johns, Bruce Nauman, Avigdor Arikha), musicians (Philip Glass, Heinz Holliger, Mascual Dusapin), philosophers (Jacques Derrida, Gilles Deleuze, and Michel Foucault), and cultural and theoretical critics (Felix Guattari, Theodor Adorno, and Maurice Blanchot). Because Beckett's influence traverses disciplinary boundaries, scholarly possibilities are virtually without limit. "New Interpretations of Beckett" will be a forum for new critical discourses on Beckett and his ongoing interdisciplinary legacy.

"New Interpretations of Beckett in the Twenty-First Century" invites work that reconnects Beckett with his own cultural and historical situation. The importance of archival access to unpublished Beckett material, the impact of the publication of *The Letters of Samuel Beckett*, and a gestational period since the official biography appeared, all lead to the next phase of Beckett Studies brimming with exciting possibilities for interpretation and evaluation. Along with recovering from its ahistorical phase, Beckett criticism is also beginning to open up new avenues of critique across the four genres in which Beckett wrote (fiction, drama, poetry, critical essay). "New Interpretations of Beckett in the Twenty-First Century" invites scholarly proposals that feature Beckett's work and/or his influence or cross-discourse with other creative artists, thinkers, or movements.

Copyright Permissions

INTRODUCTION

Beckett in History, Memory, Archive

Seán Kennedy

S amuel Beckett's writings can seem particularly resistant to historical readings, an observation that is especially true of those written
after the Second World War. Set at an anonymous crossroads, or in
strange, unfamiliar cities, the major works that secured Beckett's reputation
give the distinct impression that they are set "both anywhere and nowhere."
Accordingly, when critics first began to account for the etiolated spaces of
the mature work they concluded that Beckett was, above all, a philosopher
of being; his mature vision a deracinated exploration of what he termed "the
issueless predicament of existence."[1] In this dominant critical paradigm,
issues of historical or political interest were often ignored or deemed irrelevant as critics grappled with Beckett's apparent desire to slough off the social
and political in order to express some "bare life" that underlay all.[2] As Mark
Nixon reveals here, when Beckett was asked where Joyce stood on the issue
of the Nazi treatment of Jews he replied, "Beyond it all (31)," and similar
claims were routinely made about his own political disposition. Speaking
of Jack B. Yeats in 1954, Beckett had himself suggested that "the artist who
stakes his being is from nowhere,"[3] and this bold statement set the terms of
engagement for a generation of scholars predisposed to read his own work
in philosophical terms.

Much valuable, indeed, foundational work was done in this way, but the
philosophical consensus has been challenged recently by a growing number of scholars working with archival materials or seeking to read Beckett's
work in historical context.[4] The appearance in 1996 of James Knowlson's

biography, *Damned to Fame,* gave a major fillip to anyone wishing to locate Beckett's work in history, and the combined impact of Knowlson's findings was to suggest that Beckett's self-presentation as an "artist from nowhere" could not be simply taken on trust, and should itself be historically situated.[5] Guided by this assumption, the essays in this volume seek to restore Beckett's work to its relevant historical and cultural contexts (including the history of the text as contained in the Beckettian archive). David Lloyd has recently pointed up the peculiar challenges attending any move to historicize Beckett, warning, in particular, against the impulse to render his work historical in the same sense that we now think of the writings of James Joyce. The difference for Lloyd, and it is an important one, is that Beckett "resists from the outset any reading that would seek [...] a stable cultural or political reference."[6] That said, it has long been noted that traces of history-memory appear throughout Beckett's *oeuvre,* and their significance has yet to be fully accounted for. Very often they have been dismissed as a patina or splash of local color.[7] Here they are taken as evidence of Beckett's ongoing engagement with a range of political issues, including Nazi ideology, the Holocaust and its aftermath, and the birth of the Irish Free State. If Beckett has often been characterized as an apolitical writer—the *eminence grise* of the poststructuralist mood—this volume recovers a politically engaged writer who fully internalized the more uncomfortable implications of the events of his time. And if his work might be said to "dismantle representation" or "deconstruct history," our contributors respond by asking why and to what effect.

In this vein, a number of these essays explore the extent to which Beckett's relentless deconstruction of "subjectivity" and "identity-in-language" might have occurred as an ethical response to history and not some aesthetic negation of it. While there can be no doubt that Beckett's assault on representation was informed by his readings in philosophy, these essays suggest it was also impelled by a realization of the complicity of concepts like "Identity" and "History" in the production of authoritarian master narratives and patterns of domination. Mark Nixon traces here how Beckett's travels to Germany in the 1930s brought this home with particular force, so that he found himself torn between the need for an ethical response to fascism and a desire to remain outside political turmoil. Nazism compelled a response from Beckett rather as nationalist policy in the Irish Free State had done years previously, and Nixon's study of the *German Diaries* reveals a "curious and attentive" young writer who listened for hours at a time to Nazi propaganda, carefully noting down and subverting its claims in his journal (33).

Beckett was extremely well informed about Nazism as a result of his German visit, but the trip was intended, in the first instance, as an escape from Ireland.[8] In any event, he returned home more impatient than ever with the prevailing orthodoxies of the Irish Free State. Eamon de Valera's decision to remain neutral in World War Two dismayed him (to say the least), and James McNaughton's study of the *Watt* manuscripts here reveals how he later drew on his German experiences both to satirize fascist propaganda and "critique Irish neutrality and bourgeois apathy in the face of the German threat" (51). In this way, *Watt* combines an ethical critique of fascist expansionism with an indictment of Irish neutrality as a self-serving abdication of responsibility, expressing Beckett's "growing awareness that aesthetic decisions engage the narrative challenges presented by shoddy histories and ideological propaganda" (48). According to McNaughton, *Watt* enacts this critique of propaganda aesthetically, and in this manner he traces Beckett's politics without reducing the writing to a mere demonstration of them.[9]

What emerges from these essays is a more complex picture of a "politically literate" Beckett whose work was intimately related to its historical moment, even where he worked to obscure that relationship through successive drafts (48). In his pioneering manuscript work, S.E. Gontarski suggested that Beckett was driven by an "intent of undoing" that devalued content in favor of aesthetic form,[10] and while the value of that assessment remains, one can discern another series of transactions whereby history and memory are interrogated at the same time that they are pared down through an arduous process of revision. Beckett's intent of undoing, as Gontarski also indicates, is never quite done, and, at times, even form gave way as Beckett tried to express "the expression that there is nothing to express,"[11] because form "expresses adequacy," he told Lawrence Harvey, "and so it must be broken."[12] Indeed, it is arguable that it is this keen sense of the continuing inadequacy of representation in the face of humankind's plight that sustains (if that is the word) Beckett's enduring commitment to an aesthetic of lessness. As Andrew Gibson has suggested recently, "the sense of a larger claim, beyond oneself, outside oneself, to which one can never prove adequate [...] is conceivably the mainspring of one of the greatest projects in art of the last century."[13] From this perspective, Beckett's refusal is not of politics *per se*, but rather of a set of assumptions, routinely made in political discourse, about our capacity to adequately represent ourselves and others. Leslie Hill has argued that this refusal can only sustain "a politics beyond representation," since Beckett's work constantly "refuses to be co-opted into any cause other than that of its own irreducible singularity."[14] For Peter Boxall, however, this goes too far, and he suggests that it is rather through a reading of Beckett's "delicate tracery of reference to the cultural and political landscape of Ireland and Europe" that a sense of his politics can be discerned: tracing the traces by which Beckett "simultaneously refers and resists reference" to historical events.[15]

Drawing on the theoretical work of Dominick LaCapra, Alysia Garrison provides compelling evidence of the usefulness of the latter approach here in relation to *The Unnamable*, where, she suggests, historical events are only ever presented as "the ruins of an allusion" (91). Nevertheless, she still reads the novel as "testimonial art": an unrelenting confrontation with a post-war "aporia in history" that does not narrate the experience of the Holocaust victims directly but rather brings us "into proximity with the affective and analytic fallout of trauma" (91). In doing so, she makes clear how a different reading of the residues of history in Beckett might reveal his ethical engagements. Jackie Blackman's reading of *Endgame* performs a similar task, situating the play at "a moment in history when silent images and meaningless words became the currency of catastrophe" (73). For Blackman, it was precisely Beckett's exemplary caution about representing events in Europe that elevated *Endgame* to the status of a classic in Holocaust studies, and she reads its "oblique traces" of Auschwitz as Beckett's ethical aesthetic response to a "radical crisis in witnessing" engendered by the Holocaust (73). Alongside any intent of undoing, then, we must continue to emphasize that tenacious worldly remainder in Beckett's writing that struggles to bear witness, even if, as Boxall suggests, it is an emphasis that is difficult to maintain.[16]

Starting from a deep-seated sense of inadequacy, it seems, Beckett never simply absolved himself of political or ethical responsibility. This is, perhaps, the origin of that odd "obligation to express" that he neither explained nor reneged upon even while he asserted that there was "nothing with which to express."[17] This unerring sense of responsibility has troubled many commentators for whom Beckett's writing evinces a retreat from the political,[18] because it is so clearly at odds with any sense of inevitable failure yet so obviously provides the impetus behind much of Beckett's work. "Try again. Fail again. Fail better,"[19] might serve, in this context, as a statement of Beckett's personal political philosophy: a self-defeating rebuke to the overconfident assertions of collective national identity and historical destiny that he had encountered in Ireland, Germany and elsewhere.[20] For this reading to work, of course, we must emphasize tenacity and anticipation of failure in equal measure, avoiding any conflation of quietism with indifference. As Matthew Feldman indicates here, Beckett early on adopted a stance of "agnostic quietism" that was never "passively skeptical" (184), and although the extent of Beckett's commitment to a principled humility only became clear later, during World War Two, Feldman usefully traces its origins to his extensive readings in philosophy in the 1930s, where his skepticism was first formulated in terms of the "ethical and aesthetic acceptance of a veil separating perception from truth" (197). An awareness of this predisposition—and of the enduring distrust of the uses and abuses of representation that it engendered in Beckett—would suggest that the many traces of history-memory that *do* appear in the work cannot be dismissed as incidental and might be more usefully read as a series of quiet gestures towards contexts and events that Beckett fails to "represent" yet refuses to forget.

In this context, as both Garrison and Blackman indicate, the question of testimony becomes significant, and it is a vexed one, in that Beckett's status as "witness" or "survivor" cannot simply be assumed. What exactly did Beckett witness, and what survive? Moreover, as Paul Ricoeur has remarked, testimony is "the ultimate link between imagination and memory, because the witness says 'I was part of the story. I was there,' "[21] and this is precisely what Beckett's work does not do. Indeed, Hamm's "I was never there" in *Endgame* may be read as a tacit refusal of that role.[22] Given what happened to friends like Paul Léon and Alfred Péron, Beckett was extremely reticent about his own experiences during the War, and we must be careful to respect that gesture of silence.[23] Nevertheless, the post-war writings draw us back, again and again, to scenes of catastrophe that are all the more powerful for remaining inexplicit. Here, Robert Reginio reads both *Waiting for Godot* and *Krapp's Last Tape* "in confrontation" with the question of testimony after the Holocaust (113), an emphasis that suggests a degree of circumspection about Beckett's status as witness while still enabling a reading of these texts "after Auschwitz." Drawing on the work of Jacques Derrida, Reginio focuses on the ways in which *Krapp's Last Tape* undermines the archive's function in the production "of a legible, purposeful 'History' " (111), staging "the tension between the purported ends of the archive [...] and the necessarily excessive presence of the witness" (120). Derrida characterizes the archive as a site of anxiety,[24] a bulwark against amnesia and, in the figure of Krapp, Reginio suggests, Beckett testifies to the struggles of archiving: to remember and forget, to look back and to look forward. Reginio then traces how George Tabori's 1984 rehearsal production of *Waiting for Godot* deliberately emphasized its archival aspect in order to address "the problem of testifying about catastrophes that in themselves reconfigure testimony" (120).

To read Beckett in this way is not to deny the challenges that confront anyone trying to characterize the nature of his engagement with history. In a passage of *Writing History, Writing Trauma* that was clearly written with Beckett in mind, LaCapra notes how:

Some of the most powerful forms of modern art and writing [...] seem to be traumatic writing or post-traumatic writing in closest proximity to trauma. They [...] involve the feeling of keeping faith with trauma in a manner that leads to a compulsive preoccupation with aporia, an endlessly melancholic, impossible mourning, and a resistance to working through.[25]

LaCapra's reservation about this kind of move is that it can all too easily render catastrophic events unsayable in ways that may also render them invisible, ceding the floor to more sinister initiatives. When aporia becomes routinized, he suggests, it leads to impasse: "an interminable aporia in which any process of working through the past and its historical losses is foreclosed and prematurely aborted."[26] This, conceivably, was a lesson that Beckett learned when writing *L'Innomable* (1953), which represented a kind of terminus in his writing and haunted him for many years afterwards.[27] In LaCapra's terms, Beckett's impasse was ethical as well as aesthetic in that it threatened an end to representation in both the linguistic and political senses of that word, and it may be that Beckett's turn to theatre after the *Trilogy* was an attempt to re-negotiate the terms of his fidelity to trauma in ways that allowed him to bear witness to more than his felt inadequacy as witness. As Katherine Weiss argues here, Beckett's post-war prose replicated the horrors of the War by way of a dismembering of the human body that was also a remembering of his wartime experiences in Saint-Lô, and it was only when the body was made to disappear entirely, at the end of *The Unnamable,* that the impasse was reached. Theatre, by contrast, insisted on the need for a presence, if only of bare life, and this minimal requirement seemed to liberate Beckett from the terms of his predicament in prose: "Hamm as stated. Clov as stated. Together as stated. [...] That's all I can manage, more than I could."[28] Working with what Robert Reginio terms the "utter necessity of [...] damaged bodies on the stage" (113), Beckett recovered his impetus and went on to create some of the most important theatrical works of the twentieth century.

If this return of the body pointed a way out of the "attitude of disintegration" that was hampering Beckett's efforts in prose,[29] it also offered a new way to confront recent historical events, for, as Pierre Bourdieu has shown, the body is always conditioned by its historical moment, even (perhaps especially) when the precise nature of its effects remains obscure.[30] Here, drawing on Jacques Derrida's characterization of psychoanalysis as a "new theory of the archive,"[31] Jonathan Boulter argues that "the voice of history" in Beckett's late plays "is always too painful to bear," and so the body must function as an archive, or crypt in which its own inexpressible history is contained (138). Addressing claims that Beckett's theatre functions outside history, Boulter argues instead that the "compromised" and "repeating" body of the Beckettian stage "concretizes the idea that the past is *never* effaced" at the same time that it figures the continual, *melancholic,* end of history (146). In Beckett's later theatre, Boulter suggests, history is always in search of its "lost object–the body" (138).

By the same token, the Beckettian body is often in flight from what Paul Ricoeur terms "the wounds and scars of memory,"[32] and while history and memory have long been linked in the critical imagination, Pierre Nora has famously suggested that memory has become a purely private affair. For Nora, this psychologizing of memory, first registered in the works of Freud and Proust, was accompanied by a sense of profound malaise: "a sense that everything is over and done with, that something long since begun is now complete."[33] And if Samuel Beckett might be thought of as the inheritor of an Irish variant of this predicament,[34] so too his works may be read as extended meditations on its implications for the subject of memory. Writing here in relation to Ireland, and drawing on Ricoeur's observation that "the work of memory is a kind of mourning,"[35] my own essay highlights the persistence of "embodied memory" in Beckett's *Texts for Nothing*, tracing their refusal to mourn, and hence move beyond, certain treasured memories of an Irish father/land (21). Challenging an earlier article by Jonathan Boulter which claimed that the subject of these texts is "a subject without history or memory,"[36] I suggest the *Texts for Nothing* are haunted by both, and may be read, in part, as Beckett's response to a sense of displacement in self-exile from the land of the father. All three essays, by Boulter, Kennedy and Weiss reveal, in their different ways, the extent to which the domain of embodied history/memory leaves its trace on the Beckettian subject. For Weiss, even the dismembered bodies of late prose fragments like *All Strange Away, Lessness* and *Ping* are still working through Beckett's wartime experiences, in that the slow reemergence of the body in those texts recalls the painful process of reconstruction that Beckett witnessed on his return to France.

Overall, this is a volume about history's traces—their origins and import in Beckett's work—and, with its focus on the Beckettian archive, genetic scholarship is perhaps uniquely placed to recover their genesis. There are, however, certain dangers attached to this approach, and Dirk Van Hulle counsels here against any merely retrospective reconstruction of the history of each "finished" Beckettian text. Rather, he suggests, we should read Beckett's entire *oeuvre* as both a "work in progress" and a "work in regress," using a non-teleological methodology, "sideshadowing," that refuses the implicit teleology of reconstructive readings (179). This approach not only provides a more accurate account of Beckett's composition process, as demonstrated here through Van Hulle's careful reconstruction of the genesis of *Endgame*, it also provides a methodology responsive to Derrida's observation that "archivization produces as much as it records [an] event,"[37] and capable of addressing what Van Hulle terms "the interplay of history, memory and the archive in Beckett studies" (169).

Inevitably, when one speaks of Beckett in history, memory, and archive, one is addressing a broad canvas, yet if each contributor approaches the topic in their own way, taken together, they nevertheless provide compelling evidence of the need for a final renunciation of the idea of Beckett as an apolitical writer whose imagination functioned "almost entirely outside history."[38] This remains true even if an overall sense of a "Beckettian politics" remains elusive.[39] In 1937, recently returned from Germany, Beckett politely conceded to Thomas MacGreevy that he had "no sense of history,"[40] but he was, by then, highly attuned to the manner in which it tended to proceed by exclusion and extermination. In this context, Michael Wood has recently described Beckett's work as "a kind of memorial to caution and care, dedicated to the failures and half-lives that successful nations and persons consign to the margins,"[41] and it is this carefulness about 'representation and its discontents' that most clearly demonstrates Beckett's historical awareness. In the end, it may be the very thing that led many to characterize Beckett as apolitical in the first place—his creative destruction of the various regimes of representation sustaining political discourse in the modern period—that is best able to provide us with what he himself discerned in the rubble at Saint-Lô: "an inkling of the terms in which our condition is to be thought again."[42] Certainly, as these essays make clear, he was among the first to address that project with fitting humility.

Notes

1. Samuel Beckett, *Disjecta: Miscellaneous Writings and a Dramatic Fragment*, ed. Ruby Cohn (London: Calder, 1983), 97.
2. Giorgio Agamben, *Homo Sacer: Sovereign Power and Bare Life* (Stanford: Stanford University Press, 1998). Peter Boxall suggests Beckett's value in this period was "directly related to his widely perceived ability to give aesthetic expression to a condition that precedes and underlies being in the socio-political world." See his "Introduction to 'Beckett/Aesthetics/Politics,'" in Marius Buning et al., eds. "Beckett and Religion / Beckett/Aesthetics/Politics," *Samuel Beckett Today/ Aujourd'hui* 9 (Amsterdam: Rodopi, 2000), 208.
3. Beckett, *Disjecta*, 149.
4. For some recent examples see the articles published as "Historicising Beckett" in Marius Buning et al., eds. "Historicising Beckett/Issues of Performance," *Samuel Beckett Today/Aujourd'hui* 15 (Amsterdam: Rodopi, 2005), 21–131.
5. James Knowlson, *Damned to Fame: The Life of Samuel Beckett* (London: Bloomsbury, 1996).
6. See David Lloyd, "Frames of *Referrance:* Samuel Beckett as an Irish Question," in Seán Kennedy, ed. *Beckett and Ireland* (Cambridge: Cambridge University Press, forthcoming).

7. On this issue, see Peter Boxall, "Samuel Beckett: Towards a Political Reading," *Irish Studies Review* 10.2 (2002): 163.

8. Samuel Beckett, letter to Mary Manning Howe, 13 December 1936, Harry Ransom Humanities Research Center, The University of Texas at Austin. © Samuel Beckett Estate.

9. For a statement of concern regarding the temptation to merely transpose Beckett's art into a familiar politics, thereby diminishing its "status as inscription and event," see Leslie Hill, "Beckett, Writing, Politics: Answering for Myself," in Buning et al., eds. "Beckett and Religion/Beckett/Aesthetics/Politics," 218.

10. S.E. Gontarski, *The Intent of Undoing in Samuel Beckett's Art* (Bloomington: Indiana University Press, 1985).

11. Samuel Beckett, *Proust and Three Dialogues with George Duthuit* (London: Calder, 1999), 103.

12. See "Lawrence E. Harvey on Beckett 1961–2," in *Beckett Remembering/ Remembering Beckett,* ed. James and Elizabeth Knowlson (London: Bloomsbury, 2006), 134.

13. Andrew Gibson, "Afterword: the skull the skull the skull the skull in Connemara: Beckett, Ireland and Elsewhere," in *Beckett and Ireland,* ed. Seán Kennedy (Cambridge: Cambridge University Press, forthcoming).

14. Leslie Hill, "'Up the Republic!' Beckett, Writing, Politics," *MLN* 112 (1997): 924, 910.

15. Boxall, "Samuel Beckett," 162.

16. Boxall, "Introduction," 209.

17. Beckett, *Proust,* 103.

18. For a discussion of Beckett's "il faut" in this light, see Hill, "Up the Republic!," 919–21.

19. Samuel Beckett, *Worstward Ho* (London: Calder, 1983), 7.

20. As Mark Nixon notes here, Beckett observed in his German diary that "the expressions 'historical necessity' + 'Germanic destiny' start the vomit moving upwards" (42).

21. See Richard Kearney, "Imagination, Testimony and Trust: A Dialogue with Paul Ricoeur," in Richard Kearney and Mark Dooley, eds. *Questioning Ethics: Contemporary Debates in Philosophy* (London: Routledge, 1999), 16.

22. Samuel Beckett, *The Complete Dramatic Works* (London: Faber and Faber, 1986), 128. For more on this in an Irish context, see Seán Kennedy, "Ireland/ Europe…Beckett/Beckett," in Seán Kennedy, ed. *Beckett and Ireland* (Cambridge: Cambridge University Press, forthcoming).

23. See Knowlson and Knowlson, 79.

24. Jacques Derrida, *Archive Fever: A Freudian Impression* (Chicago: Chicago University Press, 1998), 19.

25. Dominick LaCapra, *Writing History, Writing Trauma* (New York: Baltimore, Johns Hopkins University Press, 2001), 23.

26. Ibid., 46.

27. As late as 1959, during the composition of *Comment C'est,* Beckett complained that any fresh initiative in prose threatened to relapse into the voice of

L'Innomable. In a letter to Barbara Bray, 28 February 1959, The Beckett/Bray correspondence, Trinity College, Dublin MS 10948. © Samuel Beckett Estate.

28. Beckett in a letter to Alan Schneider, 29 December 1957. See *No Author Better Served: The Correspondence of Samuel Beckett and Alan Schneider,* ed. Maurice Harmon (London: Harvard University Press, 1998), 24.

29. Beckett quoted in an interview with Israel Shenker, "Moody Man of Letters," in *Samuel Beckett: The Critical Heritage,* ed. Lawrence Graver and Raymond Federman (London: Routledge, 1979), 148.

30. According to Bourdieu, "[the body] does not memorize the past, it *enacts* the past, bringing it back to life." See "Belief and the Body," in *The Logic of Practice,* trans. Robert Nice (Stanford: Stanford University Press, 1990), 73.

31. Derrida, 29.

32. Paul Ricoeur, "Memory and Forgetting," in Richard Kearney and Mark Dooley, eds. *Questioning Ethics: Contemporary Debates in Philosophy* (London: Routledge, 1999), 6.

33. Pierre Nora, "General Introduction: Between Memory and History," in *Realms of Memory: Rethinking the French Past,* vol. 1 (Columbia: Columbia University Press, 1996), 1.

34. For more on this subject, see James McNaughton's "The Politics of Aftermath: Beckett, Modernism, and the Irish Free State," in Seán Kennedy, ed. *Beckett and Ireland* (Cambridge: Cambridge University Press, forthcoming).

35. Ricoeur, 7.

36. Jonathan Boulter, "Does Mourning Require a Subject? Samuel Beckett's *Texts for Nothing,*" *Modern Fiction Studies* 50.2 (Summer 2004): 345.

37. Derrida, 16.

38. Richard Gilman, "Endgame," in Harold Bloom, ed. *Samuel Beckett's Endgame: Modern Critical Interpretations* (New York: Chelsea House, 1988), 83.

39. On this, see Boxall, "Introduction," 207.

40. Samuel Beckett, letter to Thomas MacGreevy, 4 September 1937, The Beckett/MacGreevy correspondence, Trinity College, Dublin MS 10902. © Samuel Beckett Estate.

41. Michael Wood, "Vestiges of Ireland in Beckett's Late Fiction," in Kennedy, ed., *Beckett and Ireland,* 251.

42. Samuel Beckett, "The Capital of the Ruins," in *The Complete Short Prose: 1929–1989,* ed. with an introduction and notes by S.E. Gontarski (New York: Grove Press, 1995), 278.

CHAPTER 1

Does Beckett Studies Require a Subject? Mourning Ireland in the *Texts for Nothing*

Seán Kennedy

For all the good that frequent departures out of Ireland had done
him, he might just as well have stayed there.

Samuel Beckett, *Watt*[1]

Nobody does not have history.

Gayatri Chakravorty Spivak[2]

In 1986, with the publication of Eoin O'Brien's *The Beckett Country*,
many scholars admitted surprise at the extent to which Beckett's texts
were rooted around the specific area of Dublin in which he was born.
The book presented what Beckett found to be "beautiful photos of the sea
and mountains" around South County Dublin,[3] and placed them beside
the numerous references to that landscape in Beckett's *oeuvre,* in order to
assert place, and, more importantly, Irish places as a significant feature of
the work. James Knowlson admitted in his foreword to being "startled" by
the extent to which the texts were immersed in this area, "even Beckett's
mature, more abstracted landscapes."[4] Knowlson's "even" here indicates the
broad-based assumption that Beckett's texts grew more radical and universal in scope to the extent that they became less traditional and Irish in
setting: an assumption that underpinned almost all critical work done on

Beckett in the first thirty years. In this period, when Ireland was not wholly discounted from the picture it attained only partial recognition. Ruby Cohn suggested that Beckett's setting was "vaguely" Ireland,[5] although the more commonly assumed setting was "everywhere and nowhere." The consensus was that Beckett had effectively liberated himself from his Irish cultural predicament and was facing broader issues of identity, epistemology and representation.

O'Brien's book startled proponents of this view, but it has proved enduring. After *The Beckett Country* it has not been easy to ignore Ireland's presence, but it has proven difficult to assess its significance: it is still not clear how we should account for the Irish memories of Beckett's post-war writings. Again and again, reading the works written after 1946, we encounter certain Irish images—images of the father, of walking in the Dublin hills—that recur obsessively. But what is the significance of such memories? Or, to put it slightly differently, what was it that Beckett needed from them? Paul Stewart has highlighted our continuing "need for Beckett," and the fact that philosophers as different as Gilles Deleuze and Alain Badiou have used Beckett so extensively in clarifying their own thought suggests he continues to speak in profound ways to the needs of the contemporary moment. But is there a sense in which this need for Beckett has diverted attention from what Beckett himself might have needed? And is that something we should even be concerned with? Why, having left Ireland, did Beckett return there in his writing with such regularity, and how can we marry this to Ackerley and Gontarski's claim that Ireland "disappears" as a vital concern from the later works?[6] Ackerley and Gontarski cite Jean Baudrillard on hyper reality to substantiate their claim,[7] which suggests there may be a postmodern need for Beckett that would deny the relevance of his Irish filiations, and this despite that fact of so much evidence to the contrary. Certain needs for Beckett, it seems, need Ireland to disappear (and we must add to this, of course, that there are other needs for Beckett that would be dismayed by its disappearance).

By examining the persistence of memory in the *Texts for Nothing* here, I want to highlight the dangers attached to this need to deny Beckett his Irish memories, and also to enter a plea for a continuing engagement with history, memory, and the archive in Beckett studies. I do not wish to try to reframe Beckett's entire *oeuvre* as intrinsically Irish. Rather, I want to suggest one important reason why Irish memories endure in Beckett's work, as a way of questioning certain favored assumptions about its relationship to Irish history/memory. In particular, I want to look at certain assumptions that underpin many postmodernist accounts of Beckett's writing, framing my discussion in terms of Aijaz Ahmad's distinction between what we might call the differing moods of the modern and the postmodern.

> What is new in the contemporary metropolitan philosophies [...] is that the idea of belonging is itself being abandoned as antiquated false consciousness, [...] pushing [the] experience of loss, instead, in a celebratory direction; the idea of belonging is itself now seen as bad faith, a mere "*myth* of origins."[8]

For Ahmad, many of the crises being confronted by postmodernism were already being interrogated by modernism itself but the disposition has changed: where modernism mourned, postmodernism is triumphant.

This is a useful distinction, and one that is clearly evident in a recent article on the *Texts for Nothing* by Jonathan Boulter. Boulter's piece is a forceful critique of the current emphasis on "mourning" and "trauma" in literary studies which he decries as a deeply conservative trend predicated on "nostalgia for an originary subject and scene of loss."[9] Boulter's sense of the postmodern is similar to that identified by Ahmad,[10] but, unlike Ahmad, he is broadly sympathetic to the postmodern turn and his reading makes a number of claims consonant with a postmodern critique. He suggests, for example, that the *Texts* are spoken by "a subject without history or memory," and that they reveal trauma and mourning to be "deeply nostalgic concepts and processes that presuppose categories (self, history, memory) that themselves no longer have any operational viability."[11] Highlighting the fact that Beckett described the writing of the *Texts* as "an attempt to get out of the attitude of disintegration [of the *Trilogy* that] failed," Boulter reads them as instances of "contextual mourning," "a process of mourning within the immediate context of Beckett's oeuvre."[12] For Boulter, and it is an intriguing suggestion, these texts mourn their own inability to end, their failure to supersede the impasse at the end of *The Unnamable*.

Boulter's reading is both sophisticated and suggestive (it shared the 2004 Margaret Church *Modern Fiction Studies* Memorial Award), but I want to offer a different account of the *Texts for Nothing* here, one that retains mourning as a central trope and echoes Paul Ricoeur's conjecture that "the work of memory is a kind of mourning, and mourning [...] a painful exercise in memory."[13] Situating the *Texts for Nothing* against the backdrop of Beckett's decision to leave Ireland, I want to suggest that they be read, in part, as songs of self-exile; self-exile because there has been a tendency to conflate exile with self-exile in discussions of modernism even though they are markedly different experiences. In the difficult period leading up to the birth of the Irish Free State, for example, Peter Hart has found that "Protestant experiences of the revolution [...] ranged from massacre and flight to occasional inconvenience and indifference, from outraged opposition to enthusiastic engagement."[14] In such circumstances, not all departures were the same, and it is fair to say that Beckett, forsaking the relatively secure Protestant business community in Dublin in 1937, was a self-exile from Ireland, in contrast to many Protestant landlords who were ousted from Cork during the War of Independence, for example, who can properly be called exiles. When Mark Nixon suggests that Beckett's "frequent departures [from Dublin] were mostly necessary rather than voluntary,"[15] he is working with a more genteel version of necessity than that confronted by the landed classes, and we need to be mindful of the difference.

The distinction between exile and self-exile is worth maintaining because it has important implications for the exiled artists themselves. For one thing, as Ahmad suggests, self-exile entails a greater degree of choice regarding materials and audience as well as a less agonized relationship to home,[16] so that Beckett was in a stronger position to let Ireland go after he moved away in 1937. Since the publication of *The Beckett Country,* however, there has been a growing recognition of the extent to which it remained central to his work, albeit in abstracted form, and the critical significance of that choice remains to be fully apprehended. It would suggest that Beckett fits most comfortably with the body of writers Ahmad describes under the rubric modernism, itself

> framed so very largely by self-exiles and émigrés [...] who experienced "suffocation" in their own spaces of this globe, and [created] the predominant image of the modern artist who lives as a literal stranger in a foreign [...] city and who, on the one hand, uses the condition of exile as the basic metaphor for modernity and even for the human condition itself, while on the other, writing obsessively, copiously, of that very land which had been declared "suffocating."[17]

Ahmad does not mention Beckett in this context, but the resonances are striking. There can be little doubt that Beckett felt suffocated in Ireland (he once likened himself to an amphibian detained on dry land),[18] but he also continued to write about it, and there is an important sense in which his work is sustained by a refusal to mourn its loss.

In his work on mourning, Freud famously differentiates between successful and unsuccessful mourning, the one leading to a recovery of affect, the other to melancholia. According to Freud, only that which can be named can be mourned, since the process involves a painful working through of one's memories and hopes regarding a lost object. Therefore, mourning can only occur for an object we are conscious of having lost, and only if our memories are processed through a testing of reality which confirms the object no longer exists.[19] As Teresa Heffernan suggests, it is a process of "representing loss, translating it into symbolic language," so that libido may be "re-directed towards a new object."[20] By contrast, melancholia occurs in response to "an unconscious loss of a love-object." Either the melancholic does not know what has been lost, or "he knows what has been lost but not *what* it is he has lost in them."[21] Unable to work through his memories and hopes regarding this lost object, the melancholic fixates on it in a process of identification with a resulting "loss of capacity to adopt any new object." In this way, the process of mourning is arrested and the morbid "pathological disposition" that is melancholia develops.[22]

If we apply this distinction to Beckett we can see how his writing about Ireland is, at times, not so much a work of mourning as a refusal to mourn, a haunting, or failing to forget, issuing in a continuing confrontation with loss. By his own admission, certain of Beckett's memories of Ireland were "obsessional,"[23] especially those of his many walks through the Dublin hills with his father, and there is a sense in which he did not wish to let them go:

> I wish I was in Dublin with a car of my own to bring you out over the country we both love so much and where we have been so often together and where, if there is a paradise, father is striding along in his old clothes with his dog. At night, when I can't sleep, I do the old walks again and stand beside him again one Xmas morning in the fields near Glencullen, listening to the chapel bells. Or, near the end, in the lee of the rocks on top of the Three Rock, rubbing his feet with snow to try and bring back the circulation.[24]

Freud defines mourning as "the reaction to the loss of a loved person, or to the loss of some abstraction which has taken the place of one, such as one's country,"[25] and this letter reveals the manner in which Ireland was Beckett's lost father/land in this double sense: memories of the father are sustained by memories of the landscape, just as memories of the landscape are inseparable from memories of the father. To let go of the landscape is also to let go of the father, so deeply are the two enmeshed in memory. While the fact that the passage was written some twenty years after William Beckett's death suggests a refusal to mourn on Beckett's part: an ongoing need to identify with this Irish father/land. When his father died, Beckett told Thomas MacGreevy, "I can't write about him, I can only walk the fields and climb the ditches after him."[26] However, when Beckett left Ireland, and such walks became impossible, he turned to writing as a way of sustaining the identification.

By contrast the main thread in Boulter's analysis is that the narrator of the *Texts for Nothing* cannot mourn in Freud's sense since he has no memories

16 • Seán Kennedy

to work through towards a recovery of affect. Unable to achieve a symbolic sublation of the lost object, this narrator undermines the entire logic of the mourning process. And this is a positive development for Boulter, since mourning is an inherently conservative process predicated on a backward look: "the effort [is] to work through the shock (usually by putting the event into narrative form) in order to work back, to return, as it were *nostalgically*, to the originary scene of trauma." Ultimately, it is this quest for origins that Boulter distrusts, and why he characterizes trauma and the work of mourning as a kind of "lethal nostalgia."[27] That said, since "both processes require a subject,"[28] and the *Texts for Nothing* depict "a narrating subject without subjectivity,"[29] neither process usefully describes what is occurring in this instance:

> My initial question here was "How valid are the concepts of mourning and trauma when read into and against Beckett's texts?" Perhaps the larger question now should be "How valid are the concepts of mourning and trauma generally after being shown to be unworkable in the texts of Samuel Beckett?"[30]

Before we grant such a reading, it is worth asking whether Boulter is right to read Freud's concept of grief-work solely in terms of a backward look. In fact, Jacqueline Rose has recently argued that Freud's emphasis was on moving forward not backward. Writing during World War One, Rose suggests, Freud was anxious to "drive mourning away," and formulated the concept of grief-work "as a synonym for telos."[31] Rose feels that the grisly spectacle of the war made Freud overly keen to domesticate loss, "to provide a narrative for mourning," and that his insistence on "working through" was overdetermined by a drive for "political, or cultural, as much as psychic self-protection [...] in a moment of cultural dismay."[32] Accordingly, Rose challenges Freud's insistence that identification with a lost object is necessarily pathological, the conviction that "the living do not, and should not, identify with the dead."[33] For Rose, melancholy is what Freud uses to offload identification in order to free the way for mourning as a narrative of personal progress.[34] Melancholy bears the brunt of identification so that proper mourning can be achieved.

Against this, Rose asks whether identification might not be an entirely reasonable response to loss, and whether it might not be preferable to try and live with loss rather than drive it away. She suggests that "mourning no more comes to an end than history,"[35] so that the more useful question to ask is "not how mourning can be completed, but what is it that death, or remaining with death might permit."[36] Here, the hard-headed logic of

Freud's recovery of affect gives way to a much more vulnerable condition, one closer to Beckett's disposition. It was to Alan Schneider after his father's death that Beckett wrote, "I know your sorrow and I know [...] that in the very assurance of sorrow's fading there is more sorrow,"[37] an observation that mourns the efficacy of mourning itself. And in the case of his own Irish father/land, I think, Beckett resisted that process. If there is a sense in which the *Texts* constitute "a stupid old threne,"[38] they can also be read in Rose's terms as taking up the challenge of loss in order to go on (ever the ultimate Beckettian imperative). Irish landscapes persist in the work, then, partly because they sustain certain precious memories of the father.[39]

This would help to explain the spectrality of Beckett's later Irish land-scapes. Many commentators have remarked upon their ghostly character, most often to suggest how this reveals Ireland's diminishing significance. Ackerley and Gontarski describe Beckett's Ireland as a "specter with its sub-ject gone," the suggestion being that Ireland gradually "disappears" until it is no more than an "afterthought": a "dreamscape without presence."[40] However, Jacques Derrida's concept of "hauntology" allows us to character-ize the situation quite differently. The logic of Derrida's position is that the spectre often exerts an even more pressing claim in its absence, is "virtually more actual than what is so blithely called a living presence,"[41] and many of Beckett's letters reveal this dynamic with regard to the Irish father/land. In 1952, for example, he wrote to Pamela Mitchell that he had just taken a walk through Ussy-sur-Marne, but that "the real walk was elsewhere, on a screen inside."[42] Here, the French landscape of Beckett's present surround-ings is overwhelmed by an absent Irish father/land of memory, which is precisely the effect of the spectre in Derrida's reading.[43] Any account of Beckett's Ireland as "absent" fails to capture this peculiar dynamic of a lin-gering absent presence: "the persistence of a present past."[44]

Similarly, the haunting of the subject in Jacqueline Rose's formulation stems from the fact that we do not ever entirely overcome the loss of loved ones. In important ways, they remain: "How could identity free itself—would it still be an identity—of the traces of those who went before?"[45] Beckett's *Texts for Nothing* seem to rely on precisely this haunted structuring of subjectivity:

I saw me there, lurking behind the bluey veils, staring back sightlessly, at the age of twelve, because of the glass, on its pivot, because of father, if it was my father, in the bathroom, with its view of the sea, the lightships at night, the red harbour at night, if these memories concern me, at the age of twelve, or at the age of forty, for the mirror remained, my father went but the mirror remained, in which he had so greatly changed, my

mother did her hair in it, with twitching hands, in another house, with
no view of the sea [...][46]

Here, the themes of self, father, mother, landscape and loss all merge in a
poignant image, and it is clear that the narrator's identity is still being nego-
tiated in the presence of these ghosts: "How are the intervals filled between
these apparitions?"[47]

Elsewhere, we are told:

I'm going to rise and go, if it's not me it will be someone, a phantom, long
live all our phantoms, those of the dead, those of the living and those of
those who are not born. I'll follow him [...] mingle with air and earth
and dissolve, little by little in exile. Now I'm haunted, let them go, one by
one, let the last desert me and leave me empty, empty, sad and silent.[48]

This subject is spoken by its ghosts, and all the voices that exert claims on him
at various points in the *Texts* constitute elements of his personal hauntology:
"It's they murmur my name, speak to me of me, speak of a me, let them go
and speak of it to others."[49] Indeed, the writing of the *Texts* is itself a haunted
process expressing the nature of this predicament: "what's to be said of this
latest other [...] whose abandoned being we haunt."[50] All of which resonates
with Jacqueline Rose's claim that "to be a subject [...] is to be haunted," and
"[h]aunting, or being haunted, might indeed be another word for writing."[51]

These passages also make it difficult to accept any claim that the subject
of the *Texts* is without memory. It seems rather, as James Olney suggests,
that memory "is so central to what [Beckett] is doing that it becomes the
very subject of that doing."[52] A quick perusal of the *Texts* turns up memories
of, among other things, specific literary works like *Piers Plowman;* character's from *Waiting for Godot;* the Bible; astronomy; public houses; British
admirals; Royal Dynasties; streets in London; streets in Paris; a mother's
rheumatism, and, of course, an Irish father/land.[53] Indeed, almost every text
has recourse to memory:

See Mother Calvet again, creaming off the garbage before the nightmen
come. She must still be there. With her dog and her skeletal baby-buggy.
What could be more endurable? [...] There's a good memory. Mother
Calvet.

Often, as is the case here, the narrator is more troubled by the persistence of
memory than by its absence: "If only it could be wiped from knowledge."[54]
In what sense, then, is this subject without memory?

Perhaps we could argue that the subject is without verifiable or "true" memories in some objective sense, but, as Paul Ricoeur insists, "[i]f we can reproach memory with being unreliable, it is precisely because it is our one and only resource for signifying the past-character of what we declare to remember."[55] In such circumstances, simply jettisoning memory is hardly an adequate response to the genuine difficulties associated with remembering. Memories exist and, as Olney points out, it is not clear how things would be altered or improved by calling them "memories."[56] The question is again one of disposition, and Ricoeur's work stands as an impassioned plea for approaching memory in the first instance from the standpoint of its capacities. This does not mean that he is oblivious to its deficiencies, only that "we have no other resource, concerning our reference to the past, except memory itself."[57] And if we allow that the subject of the *Texts* does have memories, even "memories," much of the impetus behind Boulter's reading disappears, opening up a space (of uncertainty) where memories persist.

What is of interest in the context of Beckett's self-exile from Ireland is the extent to which these memories return to the theme of displacement:

And what if all this time I had not stirred hand or foot from the third-class waiting-room of the South-Eastern Railway Terminus [...] and were still waiting to leave, for the south-east. [...] Is it there I came to a stop, is that me still waiting there, sitting up stiff and straight on the edge of the seat, knowing the dangers of laisser-aller [...][58]

Here, the stuckness that Freud associates with melancholia occurs at a point of departure, and there is a sense of bi-location—of a body left in Ireland— but a sense too of an embodied existence elsewhere on the continent:

what's the matter with my head, I must have left it in Ireland, in a saloon, it must be still there lying on the bar, it's all it deserved. [...] Whereas to my certain knowledge I'm dead and kicking above, somewhere in Europe probably [...][59]

Given this emphasis on place (and displacement), I want to read the *Texts* in the light of recent theoretical work on the imbrication of memory and place, to suggest that the crisis being staged in the *Texts* is a crisis of what Edward Casey terms "implacement," or "being-in-place."[60] For Casey, there are "no disembodied memories of landscape,"[61] and he makes a compelling case for place's power "to tell us who and what we are in terms of where we are (as well as where we are not)."[62] "To be," for Casey, "is to be in place."[63]

In *History, Memory, Forgetting,* Paul Ricoeur draws extensively on these insights, reiterating the claim that memory is "intrinsically associated with places":

> One does not simply remember oneself, seeing, experiencing, learning; rather one recalls the situation in the world in which one has seen, experienced, learned. These situations imply one's own body and the bodies of others, lived space, and finally the horizon of the world and worlds, within which something has occurred.[64]

Body, place and memory are intimately linked in Ricoeur's analysis in ways that can be usefully applied to the situation of the *Texts,* and it is curious how little attention has been paid to the question of place in these fragmentary narratives, given that it is mentioned almost immediately:

> Suddenly, no, at long last, I couldn't any more, I couldn't go on. Someone said, You can't stay here. I couldn't stay there and I couldn't go on. I'll describe the place, that's unimportant. The top, very flat, of a mountain, no, a hill, but so wild, so wild, enough.[65]

This is an Irish place, as the mention of the "golden vale" makes clear, one that is close to the narrator's homeland:

> They wanted me to go home. My dwelling place. But for the mist, with good eyes, with a telescope, I could see it from here. [...] I had heard tell, I must have heard tell of the view, the distant sea in hammered lead, the so-called golden vale so often sung, the double valleys, the glacial loughs, the city in its haze, it was all on every tongue.[66]

And it is also another instance of the haunted Irish father/land, where memories of the father merge with memories of walking in a familiar image:

> my father took me on his knee and read me one about Joe Breem [...] I had it told to me evening after evening, the same old story I knew by heart and couldn't believe, or we walked together, hand in hand, silent, sunk in our worlds, each in his world, the hands forgotten in each other. That's how I've held out till now. And this evening again it seems to be working.[67]

These passages reveal how memory is, in fact, a precious (if at times harrowing) resource in the *Texts for Nothing:* "That's how I've held out till now." And the emphasis on held hands, so often described in Beckett, serves to underline the significance of embodiment in these poignant acts of remembering. If we suggest the narrator's predicament is "purely discursive,"[68] or "non-corporeal,"[69] we ignore the extent to which this remembering of Ireland is also a painful exercise in embodied memory: "I know that [...] I'm still [...] what is called flesh and blood somewhere above in their gonorrhoeal light."[70]

Reading the narrator's predicament as purely discursive also obscures what Edward Said terms the "unbearably historical" character of exile.[71] Granted, as I have suggested, Beckett was a self-exile from Ireland, not an exile. However, the Ireland that he yearned for only existed in the past, when his father was alive, and so his predicament resembled exile in the sense that there was no going back, either in time or space. And as time went by, and more of his family died, Beckett's sense of isolation from Ireland only increased further, as did the ghostly associations of the Irish landscape. Around the time of his brother's passing, Beckett wrote movingly of "the sound of the sea on the shore, and my father's death, and my mother's, and the going on after them."[72] But the fact is he was not able go on "after" them: to move beyond identification. Exiled from a precious past, Beckett never fully escaped (or wanted to escape) its lost landscapes of memory. In the end, it seemed, he only ever went back for funerals.[73] And I think this hauntological persistence of a lost father/land goes some way towards explaining the predicament of Beckett's narrator—"all ears for the old stories"[74]—in the *Texts for Nothing.*

Clearly, I am not suggesting that Beckett wanted to return to Ireland. He had no desire to belong in the Irish Free State.[75] Ireland was a father/land in the particular sense I have described—a haunted commingling of specters and spectral sites of memory—but not a fatherland, not a *patrie* in the political sense. In order to account for the lure of Irish memories in the *Texts for Nothing,* we need to contrast Benedict Anderson's idea of a national imaginary with that of topographical memory: the one being an imagining of nation, the other a remembering of place.[76] For many Irish Protestants, for whom participation in the Irish national imaginary was fraught, an investment in topographical memories of Ireland served to express affiliations that circumvented the national space. You could long for the place even while you rejected the nation, and despite the fact that they occupied the same territory. Writing of the Anglo-Irish in 1952, Brian Fitzgerald identified this important fissure exactly: "Ireland might not be their nation, but it was very definitely their country."[77] Page L. Dickinson, who published his memoirs in London in 1929, provides a clear example:

I love Ireland. There is hardly a mile of its coastline or hills that I have not walked. There is not a thought in me that does not want well-being for the land of my birth, yet there is no room to-day in their own land for thousands of Irishmen of similar views.[78]

Here, Dickinson is careful to restrict his claim on Ireland to an affinity with its soil. His work counters the newly imagined nation with rememberings of place that try to ignore its existence, seeking to recover an attachment to Ireland at a time before the new state consolidated its claims on the territory. In the same vein, Elizabeth Bowen recalled her childish confusion of the two terms "Ireland" and "island."[79] Naively, she felt she belonged to both and that both were identical, and the poignancy of her piece stems from the rude manner in which this identification fissured irremediably in the period leading to the birth of the Irish Free State.

When we think about all the "homeless mes and untenanted hims" of the *Texts for Nothing* in this context,[80] we see how their lostness resembles this painful dislocation of self across the times and spaces of exile. Many Protestants left Ireland in the period after Beckett was born there, and Elizabeth Hamilton, for one, sounds distinctly Beckettian when she describes a "feeling of being torn from the roots, lifted out of one's milieu, suspended in nothingness."[81] After she left home, Hamilton led a peripatetic existence remarkably like that of the quintessential Beckettian vagrant:

> Possibly I find it easier to adapt myself to different peoples because I am to some extent rootless, used, once the first seven years in Ireland had ended, to going now here, now there; to digging myself in, so to speak, only to find it is time to move on.[82]

More generally, Elizabeth Grubgeld has identified an "actual and attitudinal nomadism" in autobiographies written by displaced Irish Protestants linked to a conviction that "there never existed, nor will exist in the future, a psychological, social, and in some cases, physical locale in which they could feel truly at home."[83] She traces how images of childhood in Protestant memoirs are "often intertwined in memory with the image of a prelapsarian national life [...] in which the self was inseparable from specific local topographies,"[84] and, in an important sense, the haunted father/land of the Beckett country is a variation on this theme. Even if it is nearly always immediately rejected as impossible, and although a postmodern Beckett is meant to be above such things, there remains in Beckett a longing for belonging that is expressly linked to a more auspicious past:

> Yes my past has thrown me out, its gates have slammed behind me, or I burrowed my way out alone. [...] Ever since nothing but fantasies and

hope of a story for me somehow, of having come from somewhere and of being able to go back, or on.[85]

At the same time that it is consistently refused, the prospect of a return to the Irish father/land persists in the *Texts* as a tantalizing (im)possibility: "And if I went back to where all went out and on from there, no that would lead nowhere."[86]

To say this is not to suggest that the "I" of the Anglo-Irish memoirs and the "I" of the *Texts for Nothing* are equivalent (it would be an absurd claim).[87] By reading the *Texts* in this light I do not wish to occlude their complex interrogation of categories like "the subject" or "memory," only to suggest that the narrator's condition resonates with the predicament of these other exiles because he shares something of their difficult experience (and memory) of displacement. According to David Lloyd, history (and, we might add, memoir) is written with the aim of producing a "non-contradictory subject,"[88] and there is no sense in which we can ascribe such an aim to Beckett's *Texts for Nothing*. We might, however, think of them as different and more radical internalizations of an experience of displacement, registering it in the very structures of subjectivity, language and aesthetic form: "Leave, I was going to say leave all that. What matter who's speaking, someone said what matter who's speaking. There's going to be a departure."[89] This is very different, but it should not preclude us from reading the *Texts* against the backdrop of a broader crisis of identification among displaced Irish Protestants, many of whom also felt they might "dissolve, little by little in exile."[90] Beckett's writing may be unique, but there is ample precedent for his fraught relationship with Ireland.

Reading Beckett in this way also allows us to reframe the question of nostalgia in the *Texts*. In one sense, Beckett is perhaps the least nostalgic writer imaginable, and his work is often presented as the most austere dismantling of the logic of any hankering for lost origins, linguistic or otherwise. This, broadly, is Boulter's position. In postmodernist discourse after Derrida, as Ahmad suggests, "nostalgia" is often used in this way to indicate any susceptibility to the lures of belonging,[91] but it seems certain that when any evidence of the desire to belong in Beckett is conflated with nostalgia and decried in this unthinking manner, something of the emotional impulse behind his work is obscured. Indeed, the Irish father/land is one of the few recurring themes to escape Beckett's ironic self-canceling rhetoric, and some of the most haunting moments occur when nostalgia for that world is clearly evident (if only fitfully indulged). The late prose provides numerous well known examples, and it is only when we describe Beckett's writings in Rose's terms as "hauntings" that the full emotional impact of such passages becomes clear, as well as the extent to which they entail a difficult return to painful landscapes of memory.

Finally, reading the *Texts* in this historical context might also help us to approach the question of Irish history in Beckett's work. In a highly charged political atmosphere, part of the appeal of topographical memory for Irish Protestants was that it rendered the history of Protestant ascendancy less immediately discernible. As W.B. Stanford admitted, "in a land so full of ancient wrongs and grievances, the impersonal mountains, rivers, trees, flowers, paths and cromlechs offered no cause for reproach or remorse as ruined monasteries and burnt out Georgian mansions might do."[92] How ancient the grievances were is a matter of dispute, but the usefulness of the appeal to topographical memory is clear: it is both an investment in memory and a disavowal of history (especially the difficult history of Protestant ascendancy itself). In seeking to remember Irish landscapes in this way, was Beckett also disavowing the history of ascendancy? Is that why Irish history is largely absent from his texts? The answer is, of course, more complex than that, given the particulars of Beckett's relationship to that tradition, but without categories like "history" and "memory" such questions cannot even be formulated.

For this reason, although there can be no doubting Beckett's appeal to a radical postmodernist sensibility, I think we need to be wary of claims about the end of history or of memory because, as Ahmad suggests, they have the potential to result in a massive abdication of responsibility on the part of Western intellectuals. Of course the West may be all too eager to forget itself, to erase its own invidious histories and memories, but for subjects working and living in marginal or emergent locations such categories remain indispensable, such that any celebration of their loss is already an index of privilege.[93] Notoriously difficult and problematic, these categories cannot simply be jettisoned and must remain available though subject to a continuous process of interrogation and redefinition. Gayatri Chakravorty Spivak, for example, has spoken of negotiating "the necessary and impossible passage between memory and history," insisting that there can be no room for a straightforward celebration-in-loss of both categories.[94] What is required, she claims, is "a broader take on the staging of time—weaving memory and history together" to reveal how "individual memory opens into a larger history than its own."[95] Boulter's account of a Beckettian "subject without history or memory" is problematic because it cannot facilitate such a process.

Reading Beckett beyond history and memory also has implications for Beckett studies more generally, implications that sit uncomfortably with Ahmad's analysis of Western privilege. As Steven Earnshaw suggests, if one of the defining features of postmodernism has been a rejection of "history," another has been its failure to historicize its own geographical and historical position, positing itself outside history as "a synchronic and autonomous discourse."[96] The same might be said of Beckett studies, which has tended to read Beckett's work outside of any historical terms of reference. For Ahmad, this eagerness to merely dispense with the subject of history can only be a debilitating gesture, and before we dispense with the subject in Beckett studies certain questions must be confronted: What do Beckett studies renounce if they renounce history/memory? In the absence of history/memory, is an ethical Beckett possible? Does Beckett studies require a subject, and does that subject require a politics? In seeking to offer an alternative reading of the *Texts for Nothing*, then, I have also been making the case for a historicized reading of Beckett's work, on the basis that its many political resonances can only be recovered through a sustained interrogation of its particular relation to its own history/memory. As the essays in this volume combine to suggest, Spivak's call for a difficult weaving together of history and memory after postmodernism is one of the most important and challenging tasks confronting Beckett studies today.

Notes

1. Samuel Beckett, *Watt* (London: Calder, 1976), 249.
2. Gayatri Chakravorty Spivak, "The Staging of Time in *Heremakhonon*," in ed. Teresa Heffernan and Jill Didur, *Cultural Studies* 17.1 (2003): 90.
3. Samuel Beckett, letter to Mary Manning Howe, 3 September 1987, Harry Ransom Humanities Research Center, The University of Texas at Austin. © Samuel Beckett Estate.
4. James Knowlson, "Foreword" to *The Beckett Country*, by Eoin O'Brien (Dublin: Black Cat Press, 1986), xvi.
5. Ruby Cohn, *Back to Beckett* (Princeton: Princeton University Press, 1976), 63.
6. C.J. Ackerley and S.E. Gontarski, *The Grove Companion to Samuel Beckett* (New York: Grove Press, 2004), xv.
7. Ibid., xiv.
8. Aijaz Ahmad, *In Theory: Classes, Nations, Literatures* (London: Verso, 1992), 129.
9. Jonathan Boulter, "Does Mourning Require a Subject? Samuel Beckett's *Texts for Nothing*," *Modern Fiction Studies* 50.2 (2004): 333.
10. He suggests that "postmodernism celebrates the shattering and loss of traditional metaphysical and ontological categories such as truth, ethics, and the subject (and conversely, [...] modernism *mourns* the loss of these categories)," 332–33.
11. Ibid., 345.
12. Ibid., 337.

13. Paul Ricoeur, "Memory and Forgetting," in *Questioning Ethics: Contemporary Debates in Philosophy,* ed. Richard Kearney and Mark Dooley (London: Routledge, 1999), 7.
14. Peter Hart, "The Protestant Experience of Revolution in Southern Ireland," in *Unionism in Modern Ireland: New Perspectives on Politics and Culture,* ed. Richard English and Graham Walker (London: Macmillan, 1996), 81.
15. Mark Nixon, "'What a Tourist I Must Have Been': The German Diaries of Samuel Beckett" (Unpublished Ph.D. Thesis, University of Reading: 2005), 7.
16. Ahmad, *In Theory,* 131–32.
17. Ibid., 134.
18. Samuel Beckett, letter to Thomas MacGreevy, 5 August 1938, Trinity College, Dublin MS 10902. © Samuel Beckett Estate.
19. Sigmund Freud, "Mourning and Melancholia," in *On Metapsychology: The Theory of Psychoanalysis,* ed. Angela Richards (Harmondsworth: Penguin, 1991), 253.
20. Teresa Heffernan, "*Beloved* and the Problem of Mourning," *Studies in the Novel* 30.4 (Winter 1998): 559.
21. Freud, "Mourning and Melancholia," 254.
22. Ibid., 252.
23. James Knowlson, *Damned to Fame: The Life of Samuel Beckett* (London: Bloomsbury, 1996), xxi.
24. Samuel Beckett, letter to Susan Manning, 21 May 1955, Harry Ransom Humanities Research Center, The University of Texas at Austin. © Samuel Beckett Estate.
25. Freud, "Mourning and Melancholia," 253.
26. Beckett, Beckett/MacGreevy letters, 2 July 1933.
27. Boulter, "Does Mourning," 335.
28. Ibid., 336.
29. Ibid., 337.
30. Ibid., 346.
31. Jacqueline Rose, "Virginia Woolf and the Death of Modernism," in *On Not Being Able to Sleep: Psychoanalysis and the Modern World* (London: Vintage, 2004), 86.
32. Ibid., 73.
33. Ibid., 76–77.
34. Ibid., 76.
35. Ibid., 86.
36. Ibid., 80.
37. Quoted in *No Author Better Served: the Correspondence of Samuel Beckett and Alan Schneider,* ed. Maurice Harmon (London: Harvard University Press: 1998), 142.
38. Samuel Beckett, *The Complete Short Prose 1929–1989,* ed. with an introduction and notes by S.E. Gontarski (New York: Grove Press, 1995), 131.
39. Writing of the 1930s, Mark Nixon has suggested that "however painful the memories [of the father] proved to be for Beckett, there is also a distinct sense

in which the remembering moment is one of consolation, or even happiness." Nixon, "What a Tourist," 46. The *Texts for Nothing* suggest that this dynamic persisted for many years afterwards.

40. Ackerley and Gontarski, *Grove Companion*, xiv–xv.
41. Jacques Derrida, "Spectres of Marx," *New Left Review* 1.205 (May–June 1994): 32.
42. Quoted in Nixon, "What a Tourist," 46.
43. "The specter is also, among other things, what one imagines, what one thinks one sees and which one projects—on an imaginary screen where there is nothing to see." Jacques Derrida, *Specters of Marx: The State of the Debt, the Work of Mourning, and the New International* (New York: Routledge, 1994), 100–101.
44. Ibid., 101. Recently, writing with Anthony Uhlmann, S.E. Gontarski has moved closer to this latter position with the elaboration of the idea of the after-image in Beckett: "the afterimage [...] is something that lingers, haunting, no longer there but all the more there in not quite being absent." See "Afterimages: Introducing Beckett's Ghosts," in *Beckett after Beckett,* ed. S.E. Gontarski and Anthony Uhlmann (Tallahassee: University of Florida Press, 2006), 4.
45. Rose, "Virginia Woolf," 87.
46. Beckett, *Complete Short Prose,* 124.
47. Ibid., 122.
48. Ibid., 120.
49. Ibid.
50. Ibid., 150.
51. Rose, "Virginia Woolf," 87.
52. James Olney, *Memory and Narrative: The Weaving of Life-Writing* (London: Chicago University Press, 1998), 353.
53. Beckett, *Complete Short Prose,* 107; 118; 124; 102; 112; 111; 112; 122; 134; 124.
54. Ibid., 106.
55. Paul Ricoeur, *History, Memory, Forgetting* (Chicago: Chicago University Press, 2006), 21.
56. Olney, *Memory and Narrative,* 354.
57. Ricoeur, *History, Memory, Forgetting,* 21.
58. Beckett, *Complete Short Prose,* 129.
59. Ibid., 133.
60. Edward Casey, *Getting Back into Place: Toward a Renewed Understanding of the Place-World* (Bloomington: Indiana University Press, 1993), 45.
61. Ibid., 31.
62. Ibid., *xv.*
63. Ibid., 16.
64. Ricoeur, *History, Memory, Forgetting,* 41; 37.
65. Beckett, *Complete Short Prose,* 100.
66. Ibid., 101.
67. Ibid., 103.
68. Boulter, "Does Mourning," 339.

69. Jonathan Boulter, "Writing Guilt: Haruki Murakami and the Archives of National Mourning," *English Studies in Canada* 32.1 (March 2006): 126.

70. Beckett, *Complete Short Prose,* 147.

71. Edward Said, *Reflections on Exile and Other Essays* (Cambridge, MA: Harvard University Press, 2000), 174.

72. Quoted in Knowlson, *Damned to Fame,* 402.

73. "My positively last time in Dublin/Wicklow was for a few days in 68," he told Mary Manning Howe, "Peggy Beckett's funeral at Redford." Beckett, Beckett/Manning Howe letters, 7 April 1986. © Samuel Beckett Estate.

74. Beckett, *Complete Short Prose,* 103.

75. Beckett wrote to Susan Manning that "[t]o be in the streets of Paris is for me to feel how much I need France and the French way of life and how utterly impossible it would be for me to live in Ireland. I hope I shall never have to return there." Beckett, Beckett/Manning letters, 10 September 1950. © Samuel Beckett Estate.

76. Benedict Anderson, *Imagined Communities: Reflections on the Origin and Spread of Nationalism* (London: Verso, 1983).

77. Quoted in Elizabeth Bowen, *The Mulberry Tree* (London: Vintage, 1999), 175.

78. Page L. Dickinson, *The Dublin of Yesterday* (London: Metheun, 1929), 2.

79. Elizabeth Bowen, *Seven Winters* (London: Longmans, 1943), 15.

80. Beckett, *Complete Short Prose,* 150.

81. Elizabeth Hamilton, *An Irish Childhood* (London: Chatto and Windus, 1963), 146.

82. Ibid., 75.

83. Elizabeth Grubgeld, "Class, Gender and the Forms of Narrative: The Autobiographies of Anglo-Irish Women," in *Representing Ireland: Gender, Class and Nationality,* ed. Susan Shaw Sailer (Gainesville: Florida University Press, 1997), 140–141.

84. Elizabeth Grubgeld, "Anglo-Irish Autobiography and the Genealogical Mandate," *Eire/Ireland* 32.4 and 33.1/2 (1997–1998): 108.

85. Beckett, *Complete Short Prose,* 132.

86. Ibid., 111. Beckett rejected the prospect in stark terms in a letter to Mary Manning Howe: "I have no intention of going home, its not home, there is none." Beckett, Beckett/Manning Howe letters, 4 July 1955. © Samuel Beckett Estate.

87. "None of this bears the slightest resemblance to an autobiography or a confession" says Pascale Casanova, "[i]t is simply the avowal of a writing that refuses the imperatives of realism in order to mark out memories, voices from the past, childhood, shades of the mother and father, images come to haunt the memory." *Samuel Beckett: Anatomy of a Literary Revolution* (New York: Verso, 2006), 89.

88. David Lloyd, *Ireland after History* (Cork: Cork University Press, 2000), 61.

89. Beckett, *Complete Short Prose,* 109.

90. Ibid., 120.

91. Ahmad, *In Theory,* 129.

92. W.B. Stanford, *Memoirs* (Dublin: Hinds, 2001), 4.

93. Ahmad points out how "those who live [...] in places where a majority of the population has been denied access to such benefits of 'modernity' as hospitals or better health insurance or even basic literacy can hardly afford the terms of such thought." Ahmad, *In Theory*, 69.

94. Spivak, "The Staging of Time," 92.

95. Ibid., 91.

96. Steven Earnshaw, *The Direction of Literary Theory* (London: Macmillan, 1996), 77.

CHAPTER 2

Between Gospel and Prohibition: Beckett in Nazi Germany 1936–1937

Mark Nixon

In a letter of April 1967 to Kay Boyle, Beckett offered the following opinion of James Joyce's political sentiments:

> Paul Léon a Jew was [Joyce's] closest collaborator here for many years. And all the Jews he used his influence to help out of Nazi Germany when Léger was at the *Affairs Etrangères*. What was he "against"? And what "for"? Beyond it all.[1]

Beckett's work has similarly been seen as occupying a space apart from historical and political contexts, even though James Knowlson's biography *Damned to Fame* clearly demonstrated the writer's personal political awareness and engagement. Yet up until the 1990s, Beckett in the eyes of most critics and commentators was a homeless, stateless writer, who shunned geo-political problems and specificity, creating fictional worlds in order to examine the universal nature of human existence. Since then, however, critical discourse has sought to anchor Beckett's texts within recognizable geographical and historical contexts.[2]

There is much evidence to suggest that Beckett himself would have frowned upon such an approach. It was, after all, precisely this topic that led to a difference of opinion between Beckett and Thomas MacGreevy in 1938, a rift from which their friendship never really recovered. The reason for the disagreement was MacGreevy's study of the painter Jack B. Yeats, in which he attempted to characterize the painter as a specifically Irish artist. Beckett, in a long letter, told MacGreevy that he could not understand a concept such as "the Irish people," preferring the first part of the book "with its real and radiant individuals" to the second part, which focused on "our national scene."[3]

Beckett's criticism here corresponds to the approach of his early work of the 1930s, which is influenced by Joyce and a Paris aesthetic that on the whole dismissed direct political (and in particular, national) references. Even in Beckett's correspondence of the years 1929–1936 there are not many comments on political events. And these comments tend to be directed towards cultural politics, such as the Irish Censorship laws, which then found a further outlet in articles such as "Recent Irish Poetry" or "Censorship in the Saorstat."[4]

Yet if the correspondence and publications tend to avoid political and historical references, Beckett's notebooks from the 1930s paint a different picture. A set of notes dating probably from 1934 reveal that Beckett had been toying with the idea of writing a text entitled "Trueborn Jackeen" based on the history of Ireland (TCD MS 10971/2). Beckett's engagement with historical narratives is further evident in reading notes taken from Albert Sorel's *Europe et la révolution française,* George Peabody Gooch's *Germany and the French Revolution* and, perhaps less seriously, a children's book on the history of France.[5] Nevertheless, Beckett would emphasize in a letter to MacGreevy that he had "no sense of history."[6] But when Beckett embarked on a six month journey through Germany and arrived in Hamburg by boat in October 1936, he entered a country in which the angel of history was at its most wrathful. This chapter will examine the effect Beckett's journey through Germany from October 1936 to March 1937 had on his concept of history, and the way he reacted to the situation within Nazi Germany.

The political situation within Germany in late 1936 had changed radically since Beckett's five visits to his relatives, the Sinclairs, in Kassel between Christmas 1928 and Christmas 1931. In the intervening years the wider intentions of the Nazi regime had started to become apparent. By September 1936, Germany had reoccupied the demilitarized Rhineland (in March 1936), and had more or less established a complete stranglehold on the political, social and cultural spheres within its borders. In other words, there were few places in which the liberal, Weimar Germany could still be found, even in Hamburg, which Beckett described as an "island" where he found an "energetic underground of painters."[7]

Beckett traveled to Germany with some understanding of the political situation within the country. He was undoubtedly aware of the anti-Semitism propagated by the Nazis even before they came to power in 1933 through his Jewish uncle Boss Sinclair, who for that very reason left Germany ("with not much more than pyjamas & toothbrushes") when it "got too hot" for him.[8] Perhaps as a result of this, Beckett included a "Nazi with his head in a clamp" in the (unpublished) short story "Echo's Bones," written in November 1933. And in a letter to A.J. Leventhal of May 1934 Beckett punned on Hitler's *Mein Kampf* by writing "Mein Krampf [My Cramp]."[9] Furthermore, the rise of Nazism in the ensuing years, and with it the threat to individual freedom and of geographic expansion, was widely reported in the press.[10] Barely a week after his arrival in Germany, he acknowledged in his diary that he was traveling through a country that might well be at war in the near future: "They must fight soon (or burst)."[11]

During the first two weeks of his stay in Hamburg, Beckett tended to counter the harsh realities of Nazi Germany with humor. His extensive, daily notation of what he had done and experienced often incorporates play on words and satirical asides, which tend to establish a distance to his surroundings. An excursion with the owner of his "Pension," Hoppe, is thus for example described humorously despite the use of politically charged terminology such as "Anschluss":

> into a colossal modern building, then out again in company of a Herr Schlüter in green plus fours + hatchet face into another colossal modern building close by when eventually after much HH [Heil Hitler] Anschluss with Herr Dr. Reichert[12]

A similar humorous, subversive treatment of political discourse occurs when Beckett listens to an "[i]nterminable harangue by Goering on Vierjahresplan [Four Year Plan] relayed from Berlin. Sehr volkstümlich. Kolonien, Rohstoffe, Fettwaren [Very traditional. Colonies, raw materials, fats]."[13] Beckett follows this up with an entry in the *Whoroscope* notebook, which reads "Bierjahresplan" [beer-year-plan] rather than "Vierjahresplan."[14] Beckett's verbal subversion reflects his tendency to subtly undermine or satirize National Socialism in the pages of his diary, as well as in daily situations. Early on during his stay in Hamburg, he attended a charity event for German exiles in Spain, which featured a "SS Blasekapelle [brass band], bit of documentary film (Moskau droht [Moscow threatens]), speech from one Lorenz (I stretched out the wrong arm to Horst Wessel + Haydn), then more blasts from the Kapelle."[15] There are numerous such instances recorded in the diaries, such as the description of a newspaper article "with excellent photo of flight near the War memorial, Adolf Hitler Platz, running down to water (or sailing up from it). Indeed an exquisite flight."[16]

On the whole, Beckett was curious and attentive to his surroundings, taking the time to keep up with the main political events as reported in

German newspapers and on the radio. He thus shows considerable endurance (and a certain level of interest) when he listens

> like a fool to 2 hours of Hitler + an hour of Goering (opening of Reichstag, Goering reelected President, laws controlling 4 years plan extended for another 4[)], the usual from A.H. with announcement of a 20 yr. plan for development of Berlin, "reply" to Eden consisting mainly in repeated assertion that Germany's policy is not one of isolation[17]

Whilst Beckett rarely evaluates or comments explicitly upon events going on around him, his diary entries contain shrewd and perceptive observations which offer a privileged insight into the range of opinion within German society and on the street regarding the political and cultural situation.[18] As such, Beckett's German Diaries become, essentially, historical documents. There are many observations of the reality on the streets—the *Winterhilfswerk* (tellingly, Beckett buys badges in Woolworths rather than from a Hitler Youth), *Eintopfsonntag* and so forth. These observations are complemented by Beckett's notation of conversations with those supporting the new Germany: in Munich, a typesetter expresses his admiration for the Führer and shows him the place where the insurgent Nazis were shot on 9th November 1923;[19] a fellow guest at his pension propounds Germany's right to colonies; and in Leipzig a waiter explains that "the Pelz [fur] trade has gone to hell because of Jews."[20]

In descriptions of the conversations he conducted during his stay, his diaries reveal him to have been quiet when the discussion became embroiled in politics. Naturally Beckett will also have been aware of the atmosphere of surveillance in which utterances of political opposition to the regime were not taken lightly. Keeping a low profile must have seemed most expedient, especially as a foreigner. The delicate nature of the situation was brought home to Beckett by the art collector Margaritha Durrieu, who "hint[ed] how unpleasant it could be for her + Frau Fera if I published disparagements of Germany."[21] Beckett naturally gravitated toward people who stood—albeit on the whole passively—in opposition to the Nazis. He was thus for example drawn to the eminent (but under the Nazis disgraced) art historian Will Grohmann, with whom he spoke at length about the "position of German intellectual." Grohmann told Beckett that

> it is more *interesting* to stay than to go, even if it were feasible to go. They can't control *thoughts*. [...] If [regime] breaks down it is fitting for him + his kind to be on the spot, to go under or become active again.[22]

The tense atmosphere permeating the country can be felt throughout the pages of Beckett's diaries, and many entries and observations invoke an air of menace and constriction. Accordingly, Beckett was forced to acknowledge and confront the more threatening side of National Socialism. He thus showed an awareness of the possibility of war when sketching his future plans in a letter to MacGreevy, adding, "if Europe has not been obliterated before then."[23] He was equally conscious of repression within Germany's borders. This is evident from a diary entry dated 15 October 1936: "Crawl home by Jungiusstr. [...] past Jüdischer Friedhof [Jewish cemetery] (a desolation, cf. Ruysdael's *Judenkirchhof* in Zwinger [Dresden art museum], which I wonder if by now burnt)."[24] This reference to the destruction of a piece of art is the first example of a thematic complex that intensely preoccupied Beckett during his journey through Germany: the persecution of specific artists and writers through the removal of "decadent" art from museums, prohibitions to exhibit or publish, and censorship of books, journals and newspapers.

Indeed, Beckett's engagement with the darker side of National Socialism and the political situation in Germany only really intensified when he met persecuted and marginalized painters, writers and academics in Hamburg, such as Karl Ballmer, Gretchen Wohlwill, Willem Grimm und Rosa Schapire. It is in this context important to remember that Beckett himself was of course a banned, if not persecuted, author in Ireland. This possibly explains Beckett's intense interest in the Nazis' cultural politics, which is illustrated by the fact that he copied the precise wording of the prohibition to paint or exhibit levied against the Jewish painter Gretchen Wohlwill, not only into his diary, but also into his German vocabulary notebook.[25] It appears as if Beckett was in particular interested in the way in which painters and writers dealt with the restrictions, and whether they were able to uphold their creative integrity in the face of such severe threats.

In this context, Beckett differentiated between two kinds of artists. On the one hand there were the more well-known painters, such as Emil Nolde and Karl Schmidt-Rottluff, whom he deliberately did not visit as he deemed them to be "all great proud angry poor putupons in their fastness and I can't say yessir and nosir anymore."[26] On the other hand, in Beckett's view, there were painters such as Karl Ballmer, who quietly continued to work despite being hindered by a series of repressive measures: "Mild, lost almost to point of apathy + indifference. Could not exhibit even bei [with] Gurlitt [art gallery], but says it does not matter."[27] Beckett's attitude is illuminated by a further diary entry, in which he stresses that "the artist is never comfortable by definition."[28]

Nevertheless, Beckett's lifelong dislike of any kind of censorship is most clearly formulated during his trip to Germany. Many observations and descriptions of the political reality are expressed in factual tones, evident for example in the brief reference to "[p]hotographers outside Jewish shops."[29] But when he commented upon evidence of censorship, his opinion is expressed rather more directly, and vehemently. When he heard that the art critic Max Sauerlandt's study *Kunst der letzten dreißig Jahre* had been banned, his response in the diary was "Pfui!!!!"[30]

If Beckett's diary is often marked by a level of restraint in commentary, at times his distaste for the new *Weltanschauung* is expressed more clearly, as in his relationship with his German conversation partner Claudia Asher, who had been assigned to him through the Akademische Auslandsstelle in Hamburg: having listened to her talk of "national soul, of unity + might of her country" on several occasions,[31] he remarked that "[h]er Kraft durch Freude conversation kills me."[32] Beckett's already existing distaste for Nazi Germany grew with the duration of his stay. In Braunschweig he noted: "sausages in Bierstube. HH [Heil Hitler] without ceasing. Reunion of WH [Winterhilfs] Werker. Damned again."[33] And in Berlin, having summarized the extremely long discourse of his landlord on the history of the NSDAP (Nationalsozialistische Deutsche Arbeiterpartei), he adds laconically: "I find him instructive in so far as he talks in German."[34]

Rather more instructive were Beckett's meetings with two young book-sellers (and friends), Axel Kaun and Günter Albrecht.[35] He clearly felt more at ease with these two men, who both possessed a more liberal outlook, as underlined by Beckett's description of Albrecht as "not at all a Hitler Jüngling [youth]."[36] Beckett thus endorsed Axel Kaun's measured analysis of "the new Germany as one half sentimental demagogies and one half the brilliant obscurantus of Dr G. [Goebbels]," and further noted that Kaun "deplores the failure of the Jews in exile to establish a <u>spiritual criticism</u> + the futility of their protest against the inessential."[37] In particular, Beckett learned a lot about the state of censorship in Germany through the two young men, for example, that Thomas Mann was "now (in last fortnight) definitely banned, + his German citizenship taken away, because of articles in <u>Baseler Zeitung</u>."[38] It is telling that Beckett did not return to reading Thomas Mann, whose *Buddenbrooks* he had read in 1934, or that he did not read authors such as Brecht and Döblin.[39] Beckett's reading during his journey was largely of writers judged acceptable by the Nazis, an indication of the success of the ideological and cultural program of National Socialism. Although not always by his own volition, Beckett even became acquainted with several authors who were active proponents of Nazi literature. For example, his description of works inspected at a book exhibition in Hamburg's Kunsthalle includes most of the prominent authors of the National Socialist regime: Friedrich Griese, Hans Friedrich Blunck, Gerhard Schumann, Hans Heyse, and Hans Grimm, whose novel *Volk ohne Raum* was one of the earliest books to conform to Nazi ideology and thus very influential. As Beckett noted at one point, the shift in literary sensibilities meant that "Mann's world can no longer rival Grimm's (Volk ohne Raum)."[40] Furthermore, public readings during a Book Festival in Hamburg and a series of lectures promoting "Volkhafte Dichtung der Zeit" in Berlin made Beckett aware of the instrumentalization of art. Unable to attend any of the lectures, Beckett nevertheless copied some phrases from a newspaper following a Hamburg lecture by Gerhard Schumann into his diary:

Die heilige[n] Begriffe: Führer, Bewegung, Blut u. Boden, Freiheit u. Ehre dürfen nicht dem Geschwätz der Verwandlungskünst[l]er überlassen werden, die mit der Weltanschauung des NS ein Geschäft zu machen suchen.

[The holy terms: Führer, Movement, Blood and Soil, Freedom and Honour must not be given over to the babble of the fraudsters who seek to make a business out of the world view of National Socialism]."[41]

The only book of overtly Nazi persuasion that Beckett actually purchased during his trip was Hans Pferdmenges's *Deutschlands Leben* (1930), which explicitly propounds Germany's destiny of superiority. Beckett, who had bought the book following several recommendations, quickly discerned that it "seems NS Kimmwasser [bilge]."[42] As a respite from such literature, Beckett at various points during his trip turned to non-German texts: "read [Renard's] Journal + Candide, a breath of air in the dungeon. Almost happy reading that there is still such a race in Europe."[43]

If Beckett's reading during his journey through Nazi Germany was largely dictated by censorship, his inability to study modern paintings upset him far more. As early as April 1933, the Nazis had begun the "cultural cleansing" of the visual arts, removing artists and art historians who were considered "modern" or racially "impure" from their posts in galleries, museums and academies. While he was in Germany, the Nazis stepped up their campaign against "decadent" art, and on 30 October 1936, the first of the large Museums, the Nationalgalerie in Berlin, was forced to close its contemporary rooms in the Kronprinzenpalais. This was followed in November by a ban on contemporary art criticism. Beckett's diaries and letters recount his frustrated efforts to gain access to closed wings of the public galleries. He was acutely aware of the situation, telling MacGreevy in November that "the campaign against 'Art-Bolshevism' is only just beginning," and resignedly informing Mary Manning Howe that all "the modern pictures are in the cellars."[44] There are numerous instances in the diaries where Beckett notes Nazi propaganda about how "Moskau droht [Moscow threatens]."[45] Indeed, arguably Beckett's most direct exposure to Nazi ideology came when he went to see Karl Anton's propaganda film against bolshevism, *Weisse Sklaven*

[*White Slaves*], which after 1940 was shown under the title "Rote Bestien [Red Beasts]."[46]

In terms of Nazi repression of the visual arts, it was three months after Beckett's departure from Germany that the infamous exhibition "Entartete Kunst" ("Degenerate Art") was opened in Munich on 19 July 1937, which aimed at presenting the shameful decadence of modern artists and "expose dangerous criminals" to the indignation of the public.[47] Parallel to this, the exhibition "Grosse Deutsche Kunstaustellung" with Nazi-approved art was opened on the preceding day, also in Munich, in the Haus der Deutschen Kunst. Beckett's response on reading an announcement of the approved exhibition which stated that "the period of Nolde, the Brücke, Marc, etc., has been überwunden [overcome]," is indicative of his attitude towards cultural repression: "soon I shall really begin to puke. Or go home."[48]

When Beckett wrote the essay "La peinture des van Velde ou le monde et le pantalon" shortly after the Second World War, he implicitly invoked the censoring of art that he had witnessed during his stay in Germany. In the first half of the essay, Beckett spends considerable time denying the value of any art criticism, arguing that it is incapable of expressing or evaluating an art work. In this essay, which is littered with references to people he had met and paintings he had seen in Germany, Beckett also refers to the imposition of established cultural opinions:

Ne vous approchez pas de l'art abstrait. C'est fabriqué par une bande d'escrocs et d'incapables. Ils ne sauraient faire autre chose. Ils ne savent pas dessiner.

[Do not approach abstract art. It is produced by a gang of criminals and incapables. They would not know how to do anything else. They do not know how to draw].[49]

This parodies the Nazi denouncement of modern art as "degenerate," a notion to which Beckett further alludes by stating "peinture à déformation est le refuge de tous les ratés [painting which distorts is the refuge of all failures]." Beckett also refers directly to the suppression of the artist, who had to belong to a Nazi academy in order to work: "Il lui sera peut-être bientôt interdit d'exposer, voire de travailler, s'il ne peut justifier de tant d'années d'académie [It may soon be forbidden for him to exhibit, even to work, if he cannot justify so many years of academy]."[50] Draft material in the *Watt* notebooks toward the passage dealing with the mysterious painting in Erskine's room similarly refers to the removal and destruction of "decadent" art by the Nazis. Introducing three paintings Beckett had seen in Germany, the draft text refers to the fact that "they might be anywhere now, burnt, or in a lumber-room, or sent away."[51]

Nevertheless, Beckett did manage to view numerous contemporary art works before the final curtain came down. Whereas the "Entartete Kunst" exhibition of 1937 was the high point of the cultural cleansing program of the Nazis, from 1933 onward several more anti-modernist exhibitions were staged. Beckett visited such a "Schreckenskammer des Entarteten" (Chambers of Horrors of Degenerate Art) in the Moritzburg in Halle in January 1937. The majority of the pictures in the Halle "Sonderausstellung" were transferred directly into Munich's "Entartete Kunst" exhibition in July 1937, and were subsequently sold abroad or destroyed. It was in this "excellent collection" that Beckett saw numerous paintings by the main representatives of German Expressionism, such as Franz Marc's *Tierschicksal.* Beckett gives a fascinating glimpse into the general attitude toward the "degenerate" paintings in this special exhibition by recording the comments of the custodian, a "charming old bearded Diener [servant]."[52] Thus the "Diener is very trouble[d] by some perspectives that are not alas in Nature" in Lionel Feininger, Klee "draws like a child" and Nolde's "miserable thinking" is evident in the painting *Judas before the High Priests,* with its "dreadful caricature of the Sheeny that pleases the Diener."

Beckett was also able to see modern art outside such restricted exhibitions, as certain galleries had not yet removed contemporary art from public display. This was mainly due to inconsistencies in National Socialist cultural policies in 1936–1937, which stemmed from the debate between Goebbels and Rosenberg over what constituted "degenerate" art. He thus commented on the fact that whilst he was able to view drawings from Schmidt-Rottluff and Kirchner in the Zeichnungssammlung [drawings collection] of the Kronprinzenpalais in Berlin, he was unable to see paintings by the same artists in the main collection.[53] In a letter to Günther Albrecht, Beckett memorably refers to this collection of drawings as a place "wo man die Giftmischer im Intimsten ihres Schaffens geniessen darf [where one can enjoy the mixers of poison at their most intimate creativity]."[54] Beckett was particularly attentive to the fate of Max Liebermann, whose status as either "degenerate" or healthily German was heavily debated by the cultural authorities. Whereas Liebermann's paintings had been removed from the public eye in the Kunsthalle in Hamburg, his work could still be seen in the Nationalgalerie in Berlin, much to Beckett's surprise: "On ground floor so astonished at finding [...] 5 pictures by Liebermann [...] that I leave." Beckett's note that a further sixth picture, a portrait, is missing, elucidates a general policy in Nazi condemnations of modern art: figure painting tended to be banned while landscapes were generally deemed acceptable. Liebermann's status within Nazi Germany was brought home to Beckett by a man in a small pub in Staffelstein after visiting the Wohlfahrtskirche

Vierzehnheiligen: "On my mentioning Liebermann [the man] delivers a terrific harangue against Jews, other than whom Germany has no other enemy."[55]

Beckett's sensitivity to the way the Nazis were reconfiguring the cultural landscape, present and past, is reflected in his general interest in the myth, the image that they were constructing. This is evident from the numerous quotations from leading figures in the regime such as Goebbels, Hitler or Hess in Beckett's diary, and, in particular, in a notebook Beckett generally kept for German vocabulary. It appears as if Beckett was especially attentive to the Nazis' totalizing narratives of history, which was proclaimed to such a persuasive degree that it had, by the time of his visit, already become a widely indoctrinated truth. Moreover, as James McNaughton in a perceptive essay points out, Beckett's diary writing already represents a subversion of the National Socialist totalizing discourse. As he argues, the diary is fragmented and subjective, and contains numerous lists which anticipate *Watt*.[56] Indeed, the manuscript notebooks kept during the compositional process of *Watt* contain many references to Beckett's journey and the situation within Nazi Germany, all of which were jettisoned before publication of the final text. One such rather hidden reference to the Nazi regime occurs in the lines "Cheeks still wet with weeping for the Peacemaker. Shadows falling over a large portion of the inhabited globe." The allusion becomes clear from a letter Becket wrote to George Reavey in 1938, in which he referred to hearing "Adolf the Peacemaker on the wireless last night."[57]

Whilst in Germany, Beckett frequently notes in his diary that he hears the same Nazi propaganda over and over again. After a train journey in January 1937, for example, he remarks: "Rest of conversation the usual politics, almost the same words that I have heard so often."[58] Indeed, the Nazi sentiments were drilled into Beckett from all sides so that he was in a position to discern that a "Little waiter reels out the NS Evangile with only one or two errors + omissions."[59] Beckett's correlation between Nazi discourse and biblical "truth" appears several times in the pages of his diaries; thus the art collector Ida Bienert "starts with the Nazi litany," in Erfurt Beckett has to suffer "the NS Gospel from the waiter," and in Berlin an "appallingly Nazi" man "reels of the entire Gospel, as conceived for interior & exterior."[60]

By establishing this comparison, Beckett is essentially negating the claims to truth proclaimed by Nazi discourse. In his 10 March 1935 letter to MacGreevy, a letter which contains a sustained critique of Christianity, Beckett clearly notes that he has the "least faculty or disposition for the supernatural." The Bible's legitimacy as a book of truth is most famously subverted in *Waiting for Godot* through the contradictory accounts of Christ's crucifixion: and although the story makes no sense, as Didi emphasizes, "everybody" believes it. This parallels, in Beckett's mind, the Nazis' attempt to construct a new historiography, even though, as he noted when looking at the entry to a church in Regensburg, they themselves endeavored to curtail the religious sentiments of their people: "Grüss Gott [God bless you] crossed out + replaced by Heil Hitler!!!"[61]

Just how sensitive Beckett is to the complete eradication of a previous reality by the Nazis can be seen by his attentiveness to the way that Nazis renamed streets and squares (although part of his awareness also stems from a frustration of following outdated city guides). Such changes are emphasized in diary entries, as when Beckett describes a walk through Braunschweig and reaching the square "Platz der SS!!"[62] Even here, Beckett enacts a subtle resistance to Nazi discourse: having remarked that the Judenhof [Jew's Court] in Dresden was now "rebaptised Neumarkt [New Market]!," he continued to refer to it in the following diary entries as "Judenhof."[63] Beckett's perception of the way the Nazis were rewriting history also opened the possibility of reversal. When noting one of Hitler's aphorisms inscribed on the Haus der deutschen Kunst in Munich, "Kein Volk lebt länger als die Dokumente seiner Kultur [No people lives longer than the documents of its culture]," Beckett wryly commented "[p]leasant possibilities of application."

Beckett's refusal to accept the Nazi revisionary history is further evident in his comments on literary histories. Describing a visit to the bookshop Boysens in Hamburg, he noted that "[e]verything in way of history of literature, art, m. [music], prior to Machtübernahme, disparaged."[64] In keeping with his usual distrust of Nazi propaganda, it is not surprising that Beckett proceeded to buy a German literary history by Karl Heinemann because it was "written before the Machtübernahme." Beckett made a similar decision when buying a book by one Professor Knapp on architecture, having learned that he had been relieved of his position by the Nazis: "when I hear that the author is in 'retirement,' I know I am on the right thing."[65]

Beckett's criticism of the Nazis' totalitarian discourse, as well as his resistance to accept any rationalizing system of thought was most obvious when he tried to buy a book on German history. Axel Kaun had recommended Friedrich Stieve's *Abriss der Deutschen Geschichte von 1792–1935,* but on receipt of the book, Beckett was not impressed: "Just the kind of book that I do not want. Not a Nachschlagewerk [reference book], as proudly proclaimed from wrapper, but the unity of the German Schicksal [destiny] made manifest. Tod u. Teufel [damn]!"[66] He goes on to remark in his diary that he cannot read history as if it was a novel, highlighting his perception of the difference between history as fiction and as fact. This conception of history was one that Beckett emphasized throughout the 1930s. Thus for example an entry in the *Whoroscope* notebook quotes a line from Bacon's

The Advancement of Learning, "Poesis nihil aliud est quam historiae irrita-
tio ad placitum [By poetry we understand no more than feigned history or
fable],"[67] and Beckett uses the line when criticizing MacGreevy's study of
Jack B. Yeats: "You will always, as an historian, give more credit to circumstance than I, with my less than [?] interest and belief in the <u>fable convenue</u>,
ever shall be able to."[68] Beckett's point is anticipated by a diary entry of
January 1937, as he records his explanation to Axel Kaun about why Stieve's
book is inadequate:

> I say I am not interested in a "unification" of the historical chaos any
> more than I am in the "clarification" of the individual chaos, + still less
> in the anthropomorphisation of the inhuman necessities that provoke the
> chaos. What I want is the straws, the flotsam, etc., names, dates, births +
> deaths, because that is all I can know. [...] I say the expressions "histori-
> cal necessity" + "Germanic destiny" start the vomit moving upwards.[69]

Beckett's thoughts here would have a profound influence on his aesthetics
and literary work. Already *Dream of Fair to Middling Women* had implied
that the chaos of existence was not to be subjected to rational explanation.
But from 1937 onward, Beckett's emphasis on ignorance, impotence, and
not knowing intensified. More generally, the sentiments expressed here
reflect[70] Beckett's distrust of the political and historical assertions encountered in Nazi Germany.

Beckett's textural subversion of Nazism is most noticeable in the notebook where he recorded quotations by the leading figures of the regime.
Here we find material such as Hitler's statement, made in a radio broadcast
celebrating the tenth anniversary of the establishment of the Berlin Gau,
that "Nationalsozialist ist man nicht vom Tage der Geburt an [One is not
born a National Socialist]."[71] Yet the previous entry in the notebook is, like
others made around this time, a German proverb: "Mit dem ist nicht gut
Kirschen essen [It is not good to eat cherries with him]," that is to say, "he
is not friendly." Beckett's refusal of Nazi discourse through this (most probably unconscious) interweaving of political statements and jokes or prov-
erbs eventually anticipates an end to the regime. Under a quotation from
Goebbels, Beckett noted the saying: "Alles hat ein Ende, nur die Wurst hat
zwei [Everything has an ending, only the sausage has two]."[72]

Notes

This essay expands on my earlier discussion of the topic, published as "Gospel und
Verbot: Beckett und Nazi Germany," in *Das Raubauge in der Stadt: Beckett liest*

Hamburg, ed. Michaela Giesing, Gaby Hartel and Carola Veit (Göttingen: Wallstein Verlag, 2007), 79–88; translated by Marek Kedzierski as "Ewangelia i zakaz: Beckett i nazistowskie Niemcy," in *kwartalnik artystyczny* 1/2007 (53): 69–78.

1. Samuel Beckett, letter to Kay Boyle, 12 April 1967, Harry Ransom Humanities Research Center, The University of Texas at Austin. © Samuel Beckett Estate.

2. See for example the section "Historicising Beckett," in Mathijs Engelberts et al., eds., "Historicising Beckett/Issues of Performance," *Samuel Beckett Today/ Aujourd'hui* 15 (Amsterdam: Rodopi, 2005), 21–131.

3. Letter to Thomas MacGreevy, 3 January 1938, Trinity College Library Dublin, MS 10402. © Samuel Beckett Estate.

4. Samuel Beckett, "Recent Irish Poetry," in *Disjecta: Miscellaneous Writings and a Dramatic Fragment,* ed. Ruby Cohn (London: Calder, 1983), 70–76; "Censorship in the Saorstat," in Beckett, *Disjecta,* 84–88.

5. Trinity College, Dublin, MS 10969. © Samuel Beckett Estate.

6. Beckett, Beckett/MacGreevy letters, 4 September 1937. © Samuel Beckett Estate.

7. Ibid., 28 November 1936. © Samuel Beckett Estate.

8. Ibid., 7 September 1933. © Samuel Beckett Estate.

9. Charles Prentice at Chatto's acknowledges receipt of the story in a letter to Beckett dated 10 November 1933, Beckett International Foundation Archive, UoR; letter to A.J. Leventhal of 7 May 1934, Harry Ransom Humanities Research Center, University of Texas at Austin. © Samuel Beckett Estate.

10. Cf., for example, Beckett's disparaging comments on the conductor Furtwängler's conversion to Nazism in a letter to Morris Sinclair in early 1934.

11. German Diaries, 6 October 1936, UoR. See also a line from the *Frankfurter Zeitung* that Beckett copied into his diary: "War an accelerator of historical process, what in chemistry is called a catalysator" (German Diaries, 14 March 1937). © Samuel Beckett Estate.

12. German Diaries, 12 October 1936. © Samuel Beckett Estate.

13. Ibid., 28 October 1936. © Samuel Beckett Estate.

14. UoR MS 3000, 34r. © Samuel Beckett Estate.

15. German Diaries, 11 October 1936. Two months later he was pleased to observe the irony inherent in the fact that Horst Wessel was "whelped, not least suckled" in the Judenstrasse in Berlin (German Diaries, 19 December 1936). © Samuel Beckett Estate.

16. Ibid., 21 January 1936. © Samuel Beckett Estate.

17. Ibid., 30 January 1937. © Samuel Beckett Estate.

18. Beckett also recorded snippets of information regarding the political situation, including quotes by Goebbels, Hitler and Rudolf Hess, in a German vocabulary notebook (UoR MS 5006, particularly 52v-54v). © Samuel Beckett Estate.

19. German Diaries, 5 March 1937. © Samuel Beckett Estate.

20. Ibid., 28 January 1937. © Samuel Beckett Estate.

21. Ibid., 2 December 1936. © Samuel Beckett Estate.

22. Ibid., 2 February 1937. © Samuel Beckett Estate.

23. Beckett, Beckett/MacGreevy letters, 9 October 1936. © Samuel Beckett Estate.

24. German Diaries, 15 October 1936. © Samuel Beckett Estate.

25. German Diaries, 24 November 1936; UoR MS 5006, 50v. © Samuel Beckett Estate.

26. Beckett, Beckett/MacGreevy letters, 18 January 1937. © Samuel Beckett Estate.

27. German Diaries, 26 November 1936. © Samuel Beckett Estate.

28. Ibid., 4 February 1937. © Samuel Beckett Estate.

29. Ibid., 21 January 1937. © Samuel Beckett Estate.

30. Ibid., 29 October 1936. © Samuel Beckett Estate.

31. Ibid., 19 November 1936. © Samuel Beckett Estate.

32. Ibid., 1 November, 1936. Cf. Beckett's letter of 28 November 1936 to MacGreevy: "I was invited one evening to a Hausmusik. Wolf sung by a Kraft durch Freude spinster from Austria." The expression is also noted in the *Whoroscope* notebook (UoR MS 3000, 34r). © Samuel Beckett Estate.

33. German Diaries, 5 December 1936. © Samuel Beckett Estate.

34. Ibid., 6 January, 1937. © Samuel Beckett Estate.

35. For a more detailed discussion of Beckett's meeting with Kaun and Albrecht, see the essay by the latter's brother, Klaus Albrecht: "Günter Albrecht—Samuel Beckett—Axel Kaun," trans. Mark Nixon, in *Journal of Beckett Studies* 13.2 (Spring 2004): 24–38.

36. German Diaries, 6 November 1936. Beckett similarly expressed his pleasure to find Kaun "free" from the mentality expressed in tabloid and Nazi publications (German Diaries, 11 January 1937). © Samuel Beckett Estate.

37. German Diaries, 11 January 1937. Cf. Beckett's record of a conversation with his landlord Kempt: "Curious statement that Goering also was on the side of set features + earnestness, whereas Goebbels is always smiling. The Kaun (young Germany?) view of Goebbels as the sinister controlling force, + Hitler + Goering the sentimental thunderers, strikes me also as more correct" (German Diaries, 19 January 1937). © Samuel Beckett Estate.

38. Ibid. A week later Beckett told MacGreevy that "Thomas Mann [...] has had his citizenship taken away. Heinrich [Mann] down the drain long ago" (Beckett/MacGreevy letters, 18 January 1937).

39. For a more detailed discussion of Beckett and the literary scene in Nazi Germany, see my "The *German Diaries* 1936/37: Beckett und die moderne deutsche Literatur," in *Der Unbekannte Beckett: Samuel Beckett und die deutsche Kultur,* eds. Marion Dieckmann-Fries and Therese Seidel (Frankfurt a. M.: Suhrkamp, 2005), 138–54.

40. German Diaries, 22 November 1936. © Samuel Beckett Estate.

41. Ibid., 28 October 1936. © Samuel Beckett Estate.

42. Ibid., 4 November 1936. © Samuel Beckett Estate.

43. Ibid., 7 February 1937. © Samuel Beckett Estate.

44. Beckett, Beckett/MacGreevy letters, 28 November 1936 [misdated 1937]; Beckett, Beckett/Manning Howe letters, Harry Ransom Humanities Research

Center, University of Texas at Austin, 28 November 1936 [misdated 1937]. © Samuel Beckett Estate.

45. German Diaries, 24 October and 1 November 1936. © Samuel Beckett Estate.

46. Ibid., 21 March 1937. © Samuel Beckett Estate. This film was a response to Goebbels's 1934 demand for a National Socialist "Battleship Potemkin." Interestingly, considering Beckett's application to Eisenstein in Moscow to join his film school, the Russian filmmaker responded to Goebbels in an open letter, in which he stated that National Socialism and truth are incompatible, and accuses Nazis of having no idea how to create art.

47. The "Entartete Kunst" exhibition was only the start of a wider pillage of modern art; by 1938 approximately 16000 works by some 1200 artists had been confiscated, of which most were subsequently sold abroad for hard currency or destroyed.

48. German Diaries, 15 January 1937. © Samuel Beckett Estate.

49. Beckett, *Disjecta,* 120.

50. Ibid., 121.

51. *Watt* notebook 4, 2v–3r, Harry Ransom Humanities Research Center, University of Texas at Austin. © Samuel Beckett Estate. For a discussion of this jettisoned material, see David Hayman's essay "Beckett's *Watt,* the Art-Historical Trace: An Archeological Inquest," *Journal of Beckett Studies* 13.2 (Spring 2004): 95–109.

52. German Diaries, 23 January 1937. © Samuel Beckett Estate.

53. Ibid., 23 December 1936. © Samuel Beckett Estate.

54. Letter to Günter Albrecht, 31 December 1936 (UoR). In his diary Beckett noted with regard to these drawings: "so in this form they are not poison?" (German Diaries, 19 December 1936). © Samuel Beckett Estate.

55. Ibid., 22 February 1937. © Samuel Beckett Estate.

56. James McNaughton, "Beckett, German Fascism, and History: The Futility of Protest," in Marius Buning et al., eds., "Historicising Beckett/Issues of Performance," *Samuel Beckett Today/ Aujourd'hui* 15 (Amsterdam: Rodopi, 2005), 101–16.

57. *Watt* notebook 2, 39r (Harry Ransom Humanities Research Center, University of Texas at Austin); letter to George Reavey, 27 September 1938, Harry Ransom Humanities Research Center, University of Texas at Austin. © Samuel Beckett Estate.

58. German Diaries, 21 January 1937. © Samuel Beckett Estate.

59. Ibid., 28 January 1937. © Samuel Beckett Estate.

60. German Diaries, 15 February, 24 January and 20 December 1937. © Samuel Beckett Estate.

61. Ibid., 3 March 1937. © Samuel Beckett Estate.

62. Ibid., 7 December 1936. © Samuel Beckett Estate.

63. Ibid., 31 January and 11 February 1937. © Samuel Beckett Estate.

64. Ibid., 21 October 1936. © Samuel Beckett Estate.

65. Ibid., 24 February 1937. © Samuel Beckett Estate.
66. Ibid., 15 January 1937. © Samuel Beckett Estate.
67. UoR MS 3000, 59r. © Samuel Beckett Estate.
68. Beckett, Beckett/MacGreevy letters, 31 January 1938. © Samuel Beckett Estate.
69. German Diaries, 15 January 1937. Cf. Beckett's letter to George Reavey of 20 June 1945, in which he relates the death of his friend Alfred Péron in precisely such terms: "Alfred Péron is dead. Arrested by Gestapo 1942, deported 1943, died in Switzerland, on his way home, May 1st 1945" (Harry Ransom Humanities Research Center, University of Texas at Austin). © Samuel Beckett Estate.
70. Cf. also Beckett's letter to Arland Ussher dated 6 April 1938: "Did you hear the latest definition of an Aryan? 'Il agit être blond comme Hitler, maigre comme Goering, beau comme Goebbels, viril comme Roehm—et s'appeler Rosenberg'" (Harry Ransom Humanities Research Center, University of Texas at Austin). © Samuel Beckett Estate.
71. UoR MS 5006, 52v. © Samuel Beckett Estate.
72. Ibid. © Samuel Beckett Estate.

CHAPTER 3

Beckett's "Brilliant Obscurantics": *Watt* and the Problem of Propaganda

James McNaughton

> Abstract substantives charm the critic of language in the same way as surviving species of an extinct fauna; and I long for a critic of language who would be young and strong enough to expel with a pitchfork, in a great reformation of language, all abstract substantives from language. We shall have to keep the concrete substantives though, as long as we want to hold fast to the mystical belief in the reality of the beloved world.
>
> Fritz Mauthner quoted in Mauthner's
> Critique of Language, 153.

> Less terrorized by the spectre of "formalism," historical criticism might have been less sterile; it would have understood that the specific study of forms does not in any way contradict the necessary principles of totality and History. On the contrary: the more a system is specifically defined in its forms, the more amenable it is to historical criticism. To parody a well-known saying, I shall say that a little formalism turns one away from History, but that a lot brings one back to it.
>
> Roland Barthes, "Myth Today" in *Mythologies*, 112.

Beckett's unpublished manuscripts confirm his interest in political debate and the interpretive problem that propaganda presented on the long lead up to World War Two. His notebooks of German vocabulary, studiously compiled before his half-year trip to Nazi Germany, suggest what is on his mind: the first two entries "die Aufrüstung—armament" and "die Abrüstung—disarmament" are followed by other bureaucratic, economic, and political terms, including "Verteidigung—defense, justification," and, later, such words as lorry, prison, cunning, to arrest, SS, SA, and finally, "hineinziehen in bereits—to drag into (e.g., a war)."[1] As his lengthy German diaries on the trip confirm, learning such vocabulary prepared Beckett to debate Germany's right to possess colonies; to familiarize himself with the Nazi Party's illustrated newspaper, *Illustrierter Beobachter,* as well as the newspaper of the S.S., *Schwarzen Korps;* and to monitor not only the anti-Semitism of Julius Streicher's journal *Der Stürmer,* but also Nazi revisions of history, from Friedrich Stieve's sweeping history of the German people to Hitler's rewritings of recent history in speeches.[2] We can find a politically literate Beckett in these documents. He is aware of how propaganda works—whether in seemingly benign historical narratives or the incendiary harangues over the loudspeaker. Though his style is mostly objective and documentary and though he tries to separate aesthetics from politics and history, Beckett's diaries disclose a growing awareness that aesthetic decisions engage the narrative challenges presented by shoddy histories and ideological propaganda.

In this chapter, I argue that Beckett's novel *Watt* demonstrates the outcome of this thinking. Beckett wrote the bulk of *Watt* during World War Two. When he began it in Paris in February 1941,[3] five months after he joined the Resistance, Beckett was reading Hitler's *Mein Kampf* and taking copious notes.[4] He finished *Watt* in Roussillon, a small village in Provence, where he fled after his resistance cell, Gloria, had been infiltrated. He published sections in journals, including *Irish Writing* (1951–53), and issued the book entire in 1953.[5] In the novel Beckett satirizes Watt's propagandistic psychology: he is keen to give meaning to disturbing events—any meaning—in order that he might forget them. Watt interprets Mr. Knott's big house—an establishment where human authority has been bolstered by a suspect metaphysics—with a careless ideological history of the place and comic rationalizations that help him to avoid questioning authority or resisting it. Analogous with Beckett's privately recorded concerns about propaganda leading up to the war, *Watt* presents a dark parable of the challenges of confronting institutional authority.

Provocatively, therefore, Beckett's archives can reactivate his (and our) political concerns, demonstrating how he imagined art's response to politics and providing a clearer sense of the value of form and linguistic experiment in his historical context. How we link Beckett's autobiographical material or early drafts of work and the finished text nevertheless raises theoretical questions. Some of these can be dealt with by disclaimer. For instance, we should let go of our suspicions that returning to an author's archive can only represent a belated attempt to reconstruct a unified biographical subject or reanimate authorial genius. Instead, archives can provide the historical context that an author removes from his or her work but with which the work remains in dialectic. Beckett's formal considerations in his diaries, for instance, recognize the political and cultural value that narrative forms took at this time. A more animated theoretical debate over the use of archives comes from the new formalist engagement with cultural materialism.[6] New formalist critics, a diverse group to be sure, have recently argued that archival criticism (and especially new historicism) often champions readings that magnify minor textual allusions. This "machinery of archive-churning," they suggest, serves to identify cultural ideology rather than analyze the artwork as a whole, or interrogate the historical meanings of textual forms.[7] In contrast, formalists themselves are interested in the "use" of a reference rather than merely its "mention"—a distinction Richard Strier takes from W.V. Quine's philosophy.[8]

Of course, certain critics, as demonstrated by Barthes' epigraph to this chapter, have long recognized that astute historical criticism requires understanding of generic, syntactical, and organizational meanings. Indeed just as formalists now profess that they have always been interested in historical understanding—and that formalism helps one to achieve it—so cultural materialists have acknowledged, according to Kiernan Ryan, that "[r]adical historicist criticism is undoubtedly the poorer for its reluctance to meet the complex demands of a text's diction and formal refinements; for in the end only a precise local knowledge of the literary work, acquired through a 'thick description' of decisive verbal effects, will allow the critic to determine how far the work's complicity with power truly extends."[9] Having undergone a historical turn, much modernist criticism of the last two decades reflects this maturing synthesis, without needing to be so descriptively explicit.[10] Yet the terms of these debates are worth spelling out for a number of reasons. Beckett's *Watt* plays with form, syntax, and content to suggest how modes of narration bolster or critique power and ideology. Moreover, Beckett, like many experimental modernists, uses allusions that purposefully mislead us in order to question what kinds of cultural reconstructions are legitimate or even possible in a modern society. In other words, Beckett crucially confuses the distinctions between "use" and "mention," performing, for example, the rapidity with which violent history is forgotten or misconstrued, and exploring how decontextualization itself has become an important tool of ideological persuasion.

Literary interpretation stands in for wider cultural and interpretive problems in Beckett's work, and taking us well beyond the standard debates above,
Beckett's aesthetic project even makes use of his own unpublished drafts and
the interpretive challenges of using historical and biographical archives. In
More Pricks Than Kicks, for example, Beckett alludes to moments in the
manuscript of *Dream of Fair to Middling Women* that never made it to press
during his lifetime. At the end of *Mercier and Camier,* written in French in
the late 1940s but not published until 1970, the eponymous characters see
Watt in a bar, who after creating a ruckus, shouts "Up Quin."[11] Quin was
the original name of Mr. Knott in the manuscripts of *Watt,* and before that
James Molloy.[12] Since the manuscript drafts of that book, however, have
never been published, these private games guarantee that the associative leaps
Beckett's readers make will necessarily dead-end. Perhaps that is the point.
After all, Watt's generic slogan, when separated from the referent, reveals
that slogan profits from an emotional rather than a rational relationship
to its referent. Similarly, the novel *Watt* at once invites and ridicules interpretations that pan for gold in biographical and archival sources. Beckett
provocatively names the sometime narrator of *Watt* Sam, for instance. And
he ends the book with an "Addenda"—a collection of scrap quotes—set
off with a footnote instructing readers to consider the interpretive riddle
of incorporating writing and even music that never made it into the novel
proper: "The following precious and illuminating material should be carefully studied. Only fatigue and disgust prevented its incorporation."[13] This
command to study material with which the narrator could not be bothered
simultaneously acknowledges and satirizes the archival critic. *Watt* invites
us to interpret with knowledge of authorial biography, archival scraps, and
contemporary history itself—obviously key archival sources. Yet the book so
flummoxes the attempt with a mismatch between its style and reality, and
has the evidence so misinterpreted by characters seeking comfort rather than
truth, that in attempting to deduce the book's meaning with such standard
methods we encounter the same disturbing challenges that we confront in
political and historical propaganda.

By turning "uses" into "mentions," referring to unpublished drafts, earlier works, and forgotten historical events—often in a chain of increasingly
decontextualized repetitions—Beckett forces us to recognize a chilling crisis
of interpretation. We could describe this wider crisis numerous ways: the
transformation of politics into sensational consumption, the divestiture of
historical context from objects and advertised images in advanced capitalism, the ascension of ideological histories, the "over-rapid historicization"
of the recent past and so on.[14] All of these developments of modernity, in
one way or another, can be seen to facilitate the myth-making machinery

of propaganda or, at the very least, to present us with similar interpretative challenges. Roland Barthes describes the crisis in linguistic terms, noting the widespread deployment of *"second-order semiological system[s]"*: a sign "(namely the associative total of a concept and an image)" in one system becomes the signifier in a second and is made to serve another ideological concept (or signifier).[15] Drawing from wide-ranging cultural sources, Barthes gives numerous examples of content transformed to serve another ideological purpose—from the sentence about a lion benignly pulled from Greek fable to teach grammar in a Latin primer to the propagandistic cover photo on a French weekly of a black man "in a French uniform," eyes raised, ostensibly saluting the French flag. Each example radically dehistoricizes and reframes the original context, whether the lion's magnificence or the black experience that might resist such benign assimilation into French nationality. Such myth or propaganda works by reframing an historical meaning that is never completely erased: "The meaning will be for the form," Barthes writes, "like an instantaneous reserve of history, a tamed richness, which it is possible to call and dismiss in a sort of rapid alternation: the form must constantly be able to be rooted again in the meaning and to get there what nature it needs for its nutriment; above all, it must be able to hide there. It is this constant game of hide-and-seek between the meaning and the form which defines myth."[16] Beckett's *Watt* makes use of the same process by which myth and propaganda are created—not to give "historical intention a natural justification, and mak[e] contingency appear eternal,"[17] as Barthes would define both myth and bourgeois ideology, but to show this process at work. In fact, though Beckett critiques Irish bourgeois culture in this book—and as a war book *Watt* arguably manages a critique of Irish neutrality and bourgeois apathy in the face of the German threat—his experience with German propaganda at this time provides him the most formidable examples of myth. Setting *Watt* in Ireland provides a convenient space to dislocate and evaporate obvious and recent political history—providing a second-order conceptual system—that better diagnoses the kinds of narrative and linguistic forms that propelled the rising tide of Nazism along in the first place.

The last word of the last sentence of *Watt* exemplifies excellently how Beckett can pair Irish neutrality with the finger-twiddling aspects of the novel's language to suppress the reality of war, to slyly convert such an overwhelming interpretive "use" into a mere "mention." Beckett's quirky methods of containing European war generate what might be called the propaganda of apathy, a parody of actual propaganda that at once covers up the stakes of the interpretive games in this novel and reveals them with our befuddled and empty laughter. The plot of the novel follows Watt, a "university man" with a clownish red nose and hair, who takes a train from Dublin city out near the Leopardstown racetrack to Mr. Knott's house, where Watt works as a servant. At the end of the book, having left Mr. Knott's house, Watt has disappeared, evidently failing to take the train for which he has bought a ticket, and the men at the station admire the rising sun.

> And so they stayed a little while, Mr. Case and Mr. Nolan looking at Mr. Gorman, and Mr. Gorman looking straight before him, at nothing in particular, though the sky falling to the hills, and the hills falling to the plain, made as pretty a picture, in the early morning light, as a man could hope to meet with, in a day's march.[18]

The writing here is typical of the novel: the formal gymnastics distract us from darker interpretations. This last sentence resolves the permutations of the sentence before ("Mr. Nolan looked at Mr. Case, Mr. Case at Mr. Nolan, Mr. Gorman at Mr. Case, Mr. Gorman at Mr. Nolan [...]" and on), it punctuates the structure rhythmically with the elocutionary comma, and it rhymes and slant rhymes (noLAN, GorMAN, before HIM). As the locus of interpretive meaning careens from maddening permutations of micro-details to natural abstractions that provide aesthetic solace for a man, the formal poetics highlight the artificiality of the linguistic sign and the impor-tance of historical context to interpretation. The martial implication of the last phrase, "a day's march," explains it all: war is slipped between the covers of universalized nature and the familiar metalepsis that has a day marching rather than a person. The rhetorical figure removes a soldier's perspective that in the sentence seems the most natural way to judge the landscape and weather. War in *Watt* is the necessary judgment that has been removed: the rhetorical figures elide it, the "setting" in neutral Ireland confounds it, and the strange verbal games demonstrate that language has lost its dialectic with syntax, keeping time instead with a marching tune. If *Watt* is a war book, in other words, it defines itself as such by asking why so few in the book seem to admit it. More to the point, in this book, aesthetic autonomy—language expressed as formal games and here a beautiful landscape—distracts us from or covers up historical travesty. At the same time, with the satire of the characters' interpretations and in the reader's laughter and building frustra-tion at senselessness of the formalism, the book critiques this very effect. Put a little differently, if the political value of autonomous art has been that such art requires an unbiased reflective judgment, arguably preparing the capac-ity for ideological critique or even developing notions of freedom itself,[19] here autonomy appears as a kind of determinant judgment based on false neutrality or evasion. Aesthetic autonomy and political neutrality interact in a way that calls into question the open-mindedness of each.

We do not have to wait to the end of the novel, of course, for a one-word example with the shading of recent history. *Watt* provides innumerable examples of phrases and images that stencil the outline of a disturbing history—pictures that, like propaganda, cannot be filled in because they are out of context in an Irish setting and interpreted by characters whose main objective is to rationalize upsetting experiences. Perhaps the most poignant of these occurs in the third of the four chapters, through the book's chronological end. After leaving Mr. Knott's house, Watt finds himself in a mental sanatorium or holding camp where double rows of barbed-wire fence surround a number of pavilions and "gardens." In this episode, Watt's friend Sam narrates, and his interpretation of why humans are penned like this performs the confusion that Barthes describes, where history is tamed, recalled and dismissed in quick sequence. Double rows of barbed-wire fences—indeed barbed wire itself—began to symbolize during the war totalitarian oppression and the irrational limits of man's rationality, just as a hole in such a fence became Primo Levi's famous symbol of liberty.[20] Beckett wrote a good portion of *Watt* while hiding, evidence indicates, within a 50-mile radius of at least four internment camps in the south of France.[21] Had he been caught, like many in his Parisian cell Gloria, he would have been detained in such a holding camp before being sent—though he would not know the full extent of their murderous operations until after the war—to Ravensbrück, Mauthausen, or Buchenwald further north.[22]

Beckett's text, however, removes such symbolic interpretations of wire camps to trust that they will reappear as the elephant in the room, as a guilty and mirthless laugh that obviously relies upon the reader's awareness of contemporary history. Here is Sam's interpretation:

This garden was surrounded by a high barbed wire fence, greatly in need of repair, of new wire, of fresh barbs. [...] Now converging, now diverging, these fences presented a striking irregularity of contour. No fence was party, nor any part of any fence. But their adjacence was such, at certain places, that a broad-shouldered or broad-basined man, threading these narrow straights, would have done so with great ease, and with less jeopardy to his coat, and perhaps to his trousers, sideways than frontways. For a big-bottomed man, on the contrary, or a big-bellied man, frontal motion would be an absolute necessity, if he did not wish his stomach to be perforated, or his arse, or perhaps both by a rusty barb, or by rusty barbs. [...] an obese wet-nurse, for example, would be under a similar necessity. While persons at once broad-shouldered and big-bellied, or broad-basined and big-bottomed [...] would on no account, if they were in their right senses, commit themselves to this treacherous channel, but turn about, and retrace their steps, unless they wished to be impaled, at various points at once, and perhaps bleed to death, or be eaten alive by the rats, or perish from exposure, long before their cries were heard, and still longer before the rescuers appeared, running, with the scissors, the brandy and the iodine. For were their cries not heard, then their chances of rescue were small, so vast were these gardens, and so deserted, in the ordinary way.[23]

Certainly gardens separated by two rows of fence and barbed wire constitute a safety hazard for those obliged to walk between them. Yet this interpretation is breathtakingly irrelevant. How absurdly wrong Sam is to imagine "obese wet-nurses" or the perils of navigating barbed-wire fences for the well fed, or that a team of "rescuers" is on the lookout with "brandy and iodine." Sam's rationalist outlook—that leads him to seek meaning in terms of probable causes or probable outcomes—is undone by an imaginative range of possibility evidently stuck in a prewar context with a frame of reference uncomplicated by recent history.

Yet barbed wire's more sinister history lurks in his reasoning too, and Sam's interpretation also seems like a desperate attempt to anesthetize disturbing contemporary history. By World War One, barbed wire, cheap and resistant to artillery, had become the most efficient defense against an advancing army and a key symbol of trench warfare. Sam's description of someone impaled, bleeding to death, eaten alive by rats, or perishing from exposure echoes familiar descriptions of a trench solider entangled in barbed wire. Sam tries to tame the dark contemporary reality lurking here by classifying this camp with both the benign past and natural equivalents. He admits to being "very fond of fences, of wire fences, very fond indeed [...] to all that limited motion, without limiting vision, to the ditch, the dyke, the barred window, the bog, the quicksand, the paling, I was deeply deeply attached, at the time deeply deeply attached."[24] He classifies together natural geographic and manmade technologies of incarceration, ignoring the qualitative difference between a system of human incarceration and natural geographic features. By repeating "deeply deeply attached" Sam's interpretation reveals itself as a perversion of propaganda that makes benign the uneasy history it still contains. Not only is he physically restricted by such barriers and therefore attached quite deeply to them, but trying to leave or navigate the fences could leave the barbs "deeply deeply attached" indeed. The literalness of the sentimental commonplace ironizes euphemism as a coping mechanism, a desire to self-deceive that makes the fantasy of myth so compelling.

Beckett's technique here—to satirize rhetorical tropes, clichés, and forms of rational logic that mystify and erase, make natural and eternal the radical qualitative changes in the twentieth century—raises a number of questions. First, had Beckett wanted us to consider contemporary history, would he not have written about his or others' experience in the war directly? Second, is it reasonable to assume that all readers know Beckett's biographical involvement in history or are willing to take textual hints to the archive to figure out their importance? Finally, if there is a critique of fascism at work here, does comedy, even mirthless comedy, not make light of German horrors by integrating them into a strangely benign Irish perspective? The answers begin by noting that *Watt* takes many risks to satirize the kind of thinking that neutralizes the horror of contemporary history. Foremost among them is an assumption that readers know basic twentieth century history, evidenced negatively by our ability to laugh at the main characters' misinterpretations. As for biographical evidence, by calling the book's narrator "Sam," to whom Watt purportedly though unbelievably narrates the entire book, Beckett simultaneously invites this comparison to his biography and dismisses it. Sam's positivist outlook makes him little like his authorial namesake. Yet even here the book bends the boundary between legitimate and illegitimate literary interpretation to suggest that readers need to learn the mechanics of myth, the disturbing element of the almost familiar, the erased context, the "tamed richness" of an ideological reframing of fact. Avoiding documentary descriptions of his own experience or horrific events more generally, steers the moral energy of the book from what happened—*Watt*'s key concern—to the more important questions of why it happens and what logic do interpreters use to find their neutrality, historical innocence, or consolation for inaction? If by failing to specify actual historical camps *Watt* confuses a precise analysis of their terrifying development in Germany, for instance, to see them as a Nazi phenomenon alone historicizes them too quickly.[25] The first concentration camps in Germany, Giorgio Agamben reminds us, "were the work of the Socialist Democratic Governments" used to intern communist militants and Eastern European refugees.[26] Article 48 of the Weimar constitution allowed fundamental rights to be suspended in an emergency—a state of exception the Nazis made permanent. The same historical ambiguity, of course, can be said for fascism's propaganda: Beckett himself identified the "sentimental demagogics" and "brilliant obscurantics" of Goebbels as learned from the USSR. "Certainly," he writes, "he is a pupil of their technique."[27]

From another perspective, however, it might be asked, does reading the Irish backdrop as simply a site of benign neutrality against which contemporary European history is evoked and forgotten minimize the peculiar way that the book can be investigated, as W.J. McCormack urges of Yeats's *Purgatory,* as an examination of "the place of Protestant Ascendancy ideology in the broader field of European racism."[28] McCormack's work convincingly argues that the idealization of an ascendancy tradition from the eighteenth century to the present—in big house novels and W.B. Yeats's mythologizing, for example—helped to mystify economic relations with Britain, ignore the existence of Ireland's middle-class, and romanticize class disparities. He correctly identifies Beckett's *Watt,* particularly with its satire of Yeats's *On the Boiler,* as a text that begins to undo such mystification. Indeed, scholarship of the last two decades has made much of the book's second chapter to expose Beckett's satire of both the Anglo-Irish big house novel—a form whose narrative value seems spent—and the lingering affection for the power structure itself voiced by Yeats and Elizabeth Bowen, among others. And undoubtedly, the book also uses Ireland's semi-colonial past—Watt's discussion of the "colony" of "famished" dogs, for instance—to evoke and dismiss the nineteenth century Irish famine, a historical catastrophe the causes of which Watt comically scrambles. Ireland, in other words, is not simply a blank space to analyze problems abroad.

Though these emphases are certainly important, I suggest that the evocation of Irish spaces and Watt's convoluted interpretations of the Irish past are more complex than a satire of Anglo-Irish ideology, even as Yeats's aristocratic sympathies aligned periodically with European fascism. This is especially so since by this point Anglo-Irish political power has largely collapsed, the myth of its eternal serialization has been largely destroyed, and attempts to reconstruct it appear somewhat foolish. Instead, the decaying Anglo-Irish institution more convincingly ironizes Watt's considerations of Mr. Knott's eternal power, and by extension the kinds of fascist power falsely claiming eternal legacy abroad. Yet we need not feel like we have to choose between championing the book as Irish or continental; instead, we should understand how these various readings interact. The book's use of an "Irish" setting enacts a parody of how propaganda works by swallowing contemporary history with formal games; yet the vagueness of the settings potentially undermines Irish exceptionalism. Holding camps, we might remember, were also employed by the Irish during the Irish civil war, as McCormack points out. That Ireland is nevertheless an unlikely place to examine fascism and therefore can be chosen to anesthetize camp imagery or evoke distant war by gauging the fairness of weather for "a day's march," shows that its relationship to the horrors unfolding in Europe is one of assumed exception. Ironically, this relationship of reference by exception is analogous to that which concentration camps themselves have to the law, in that they exist as spaces beyond the law, in order to guarantee the law—a key ambiguity that shows how spaces of legal exception themselves can be the constituting factor of barbarity in modernity.[29] In this sense the inter-national perspective of the book's play between a neutral country where historical content is formalized and rendered autonomous and a terrorized Europe can suggest the grounding conditions that actuate that terror.

These reasons give some further justification for looking at Beckett's diaries in Germany. Though he understood propaganda cross-culturally, the diaries nevertheless disclose how Nazi Germany provoked him to investigate how an experimental aesthetic might respond better than documentary realism to such a dangerous political ideology. Beckett's diaries from the six-month trip recognize how political propaganda works, from the mouth of Hitler to the more subtle forms of artistic and political histories of Germany. The day after Hitler's anniversary speech at the Reichstag on January 30 1937, for instance, Beckett writes in his diary that he "read more of AH's selfgratulations" and records that Hitler says of his political takeover in 1933 that "not a pane of glass was cracked (!)" and that "after 1933 the war was over, all subsequent Blutbäder to be entered under head of *evolution*."[30] He correctly reads Hitler's use of the word "evolution" as a verbal cover for violence, "massacres," as the German "Blutbäder" translates. Given Beckett's attention here to word choice that makes "contingency appear eternal," we should be unsurprised that he also fumes at the metahistorical narratives of German history written by Friedrich Stieve. This historian might be familiar for his more obvious agenda, such as the later pamphlet, "What the World Rejected: Hitler's Peace Offers 1933–1939."[31] But in 1937, when Beckett reads what might have been his *Geschichte des Deutschen Volkes* (1934), he does not comment upon Hitler's endorsement on the title page ("German People, forget 14 years of decay, rise from 2000 years of your German history"). Rather the more enduring problem that facilitates Hitler's strategic forgetting is the form. "[J]ust the kind of book that I do *not* want," Beckett writes, "[n]ot a *Nachschlagwerk* [G. reference work], as proudly proclaimed from the wrapper, but the unity of the German schicksal [G. fate] made manifest [...] What I *want* is precisely a Nachschlagewerk, and can't read history like a novel. [...] Schicksal = Zufall [G. fate = chance] for all practical human purposes."[32] Stieve's book begins with the history of the German people from 2000 years ago and ends, in the tones of *Mein Kampf,* with rhapsodic praise for the rise of Hitler, a man from the people, a stunning inevitability to celebrate.

Beckett expresses his disgust for anthropomorphized narratives, history as ideological and fatalistically sweeping ("'historical necessity'") as well as for "'Germanic Destiny,'" which the first page of *Mein Kampf* identified as the right to unify the German people and acquire colonies within Europe.[33] Beckett opposes on epistemological grounds Hitler's "essential" history—the sweeping "forces which are the causes leading to those effects which we subsequently perceive as historical events"[34]—with inessential history, a strict documentation of fact allied with an outlook of chance: "names, dates, births, + deaths, because that is all I can know." And he rejects outright Nazi attempts to unify or clarify the "individual chaos" through their shoddy logic—a kind of mystical rationalism that abstracts universal qualities from millions of individuals, and then reapplies them as the original national quality, the ordering cause. Doing so not only compels "the will of the individual to sacrifice himself for the totality,"[35] but also leads to the anti-Semitism that anthropomorphizes unity's inversion, chaos.

I have argued elsewhere[36] that Beckett's diary itself performs the debate, since he devotes his entries mostly to a compendium of facts, lists of paintings and street names, a memorial record that seems to pit Beckett's own carefully detailed "Schicksal" against narratives of German "Zufall." But this alternative of using documentary realism as a response to fascist ideology, a popular notion at the time, seemed to offer only limited critical capacity, not least in a private diary.[37] After a discussion with Axel Kaun who "deplores the failure of the Jews in exile to establish a *spiritual criticism* & the futility of their protest against the inessential," Beckett seems to recognize that a successful ideology can easily withstand factual contradiction. He admits the absurdity of his own diary, calling it merely a treatment of a content, "an 'open-mindedness' that is mindlessness."[38] Reluctant to give history rational shape, and yet certain that listing details has little chance of critique, Beckett returns to Ireland determined to devote himself to developing instead an elaborate satire of ideology, metahistory, and what Austrian Linguist Fritz Mauthner would call "word-superstition." Partly under Mauthner's influence, whom Beckett had been reading for years, he famously claims that he will "tear language apart," so that the "void may protrude like a hernia."[39]

These announcements have seemed like mere quirky aesthetic or linguistic projects designed to show that language is non-referential. Indeed, Mauthner defines word-superstition as the seductive power of language that leads us to assume that words, particularly abstract nouns, refer to a given substance, instead of what is at best a set of sensory impressions or the memory of those impressions. Yet Beckett's linguistic skepticism has a tighter political focus than Mauthner's philosophy, a philosophy that made an exception for the emotional power of abstract nouns in political and poetic discourse.[40] In *Watt,* which obviously rejects documentary realism as a form of critique, Beckett explores many methods for "tearing language apart" and with an experimental range far more complex than this deconstructive cliché implies. As we have seen, he deploys second-order sign systems—individual words, symbols, or historical narratives whose vexed histories are erased in a new context—in order to parody the process by which propaganda refashions historical meaning. Yet the book conducts a much more thoroughgoing engagement with the rational operations that disturbed Beckett in Germany, especially anthropomorphized history and mystical rationalism.

Watt's experience in Mr. Knott's house, for instance, can be read productively as a confrontation with language that mystifies power. The landlord Mr. Knott himself is a kind of "hernia" in the void; he resembles a formal allegory more than he does a realistic character. Watt rarely sees him, never face to face, and Knott's body can change endlessly—so that he appears like one of Nietzsche's idols: eternal, fixed, changeable, unknowable, and mysteriously captivating. Allegory becomes the shorthand of a bankrupt and seductive metaphysics, now the domain of narrative convenience and political power rather than truth. Watt's experience in Mr. Knott's house, should, we are told, lead him to critique abstractions, see through allegory, but, stunningly, Watt resists by generating narratives—any narratives—that make sense of the data presented to him, so that he may forget them. The targets of this critique become clear: the speciousness of transcendental notions such as "historical necessity"; providential histories that exploit and depend upon the abstraction of language, a process normalized by advertising and propaganda; and simple documentary accounts of the past that facilitate forgetting.

Although Mr. Knott's metaphysical being has sometimes been associated with the mystery of God,[41] he represents as much a political problem as a theological or philosophical one. The book cogently suggests that metaphysical meaning, when appropriated allegorically by humans or human institutions, mystifies power and entraps subjects. In Mr. Knott's establishment, Watt's servitude enacts an irrational enthrallment to the metaphysical abstraction that fascist ideology appropriates to transform its leaders into mythic figures and to narrate its past. As Watt gets closer to Mr. Knott—moving from a first-floor servant to second-floor personal servant—he undergoes sense-deprivation, a thinly veiled parable that transforming worldly categories or institutions into eternal metaphysical institutions requires severing language from its relationship to reality. In itself, this observation is hardly political. But the book specifically demonstrates how advertising and sacred language exploit and perpetuate the separation of experience from language, which ultimately destroys Watt's ability to understand class and power relations, most particularly the gross economic inequities upon which the house has been built. By specifying the discourses that mystify language and by exploring the loss of political intelligence, Beckett couples "Knott's establishment" to basic narrative components of Nazi politics, namely providence, the rejection of class-based critiques, and the justifications for imperialism.

We are tipped off early in the book that Mr. Knott might be simply a convenient metaphysical entity, a necessary cause to justify the existence of others. As one of the servants, Arsene, tells the story of two maids near the beginning of the book, he is forced to introduce another person in order to explain the existence of the maids:

> and let there exist a third person, the mistress, or the master, for without some such superior existence the existence of the house and parlour-maid, whether on the way to the house and parlour, or on the way from the house and parlour, or motionless in the house and parlour, is hardly conceivable.[42]

Like the master or mistress in this story, Mr. Knott's "superior existence" is a logical outcome: if you imagine a maid, you must assume she works at a master's house. And, by Watt's logic, since Mr. Knott's represents the agency of a longstanding institution for which he is the anthropomorphic representative, he must have abided for all time like an "oak [...] and we nest a little while in his branches."[43] Similarly Watt begins to conceive, in a parody of the Christian Liturgy's *Gloria Patri,* that the institution itself must be eternal:

> Watt had more and more the impression, as time passed, that nothing could be added to Mr. Knott's establishment, and from it nothing taken away, but that as it was now, so it had been in the beginning, and so it would remain to the end, in all essential respects.[44]

As Hesla points out, Beckett also changes the "world without end" to "to the end," an important modification that reminds us that this establishment is not actually eternal, but merely gives that "impression." This crucial distinction reiterates the point that the metaphysical authority granted by language is an illusion that mystifies power. It can be confronted and it can end.

Watt is not an ideal candidate for this project of seeing things without ideological mystification or for generating critique: Beckett makes Watt's rationalizations the object of satire for this very reason, to show how badly bourgeois man wants to delude himself. The book frames Watt's arrival to Mr. Knott's house with the expectation that he will learn to critique language and begin to see the grip of word-superstition on his perception. In the house "words [...] beg[i]n to fail," yet confronted with this disturbing experience, Watt nevertheless is compelled to rationalize it away: "the state in which Watt found himself resisted formulation in a way no state had ever done, in which Watt had ever found himself."[45] In a sense, Watt exemplifies false consciousness *par excellence*. He willingly assigns nouns and causes to events—the "what" and "how"—so that he does not actually ever need to know "why" events happen, and so that he can readily and comfortably forget them: "[T]o explain had always been to exorcise, for Watt,"[46] and "Watt's need for semantic succour was at times so great that he would set to trying names on things, and on himself, almost as a woman hats."[47] Watt trusts that language can be applied to objects, and he opposes the pressure to critique language, in part, the book suggests, to preserve his position as part of the establishment. Unconcerned with "what [events] really meant, his character was not so peculiar as all that," Watt strives for what events can be "induced to mean, with the help of a little patience, a little ingenuity."[48]

In effect, Watt should learn when working for this institution the foolishness of his intellectual compulsion; he should begin to see things for what they actually are and achieve a kind of awakening. When Watt first arrives at Mr. Knott's house, the servant whom he will replace, Arsene, describes the stages that one undergoes in the house. First, one feels indignation at having to perform "tasks of unquestionable utility," especially since before coming to the house one had considered that "to do nothing exclusively would be an act of the highest value." Yet the indignation wears away, because "he comes to understand that he is working not merely for Mr. Knott in person, and for Mr. Knott's establishment, but also, and indeed chiefly, for himself." His regrets, "lively at first, melt at last [...] into the celebrated conviction that all is well, or at least for the best." But then, one day "the horror of what has happened reduces him to the ignoble expedient of inspecting his tongue in a mirror," and a profound change comes over him.[49] The passage describes how laboring for an institution socializes dissent by supplying private gain, until one wakes up to complicity in some horror. The generalized language makes this process profoundly provocative: is Beckett explaining that most difficult of questions for anyone looking back to 1930s fascism: why did you go along with it? Or is Beckett explaining more generally how we become complicit with economic systems that perpetuate suffering because we have a stake in the status quo? Arsene's descriptions suggest that Watt's experience should lead him to address versions of these questions.

Arsene describes the radical transformation in the same terms Mauthner uses to describe the result of no longer being beholden to word-abstractions. Arsene claims that on this day of realizing the horror, he experienced the world with a "perception so sensuous" that the world "underwent an instantaneous and I venture to say radical change of appearance."[50] Looking at the sun on the wall, he perceived it "so changed that I felt I had been trans-ported, without my having remarked it, to some quite different yard, and to some quite different season, in an unfamiliar country."[51] Arsene describes perception unbiased by habitual preconceptions. Just as in Beckett's *Proust,* where stripping habit—ideological and otherwise—is the final goal for a finer perception of the object, so here Arsene has achieved perception with-out the preconception of language, without ideological mystification. Arsene describes this change of affairs as "the reversed metamorphosis. The Laurel into Daphne. The old thing where it always was, back again."[52] The result of this transformation is that the eternal concept, symbol, allegory, and word-superstition lose their abstract quality or conceptually disappear.

Nazi ideology alone does not create word-superstition, nor is it that fas-cism alone that benefits from it. Yet as others—particularly the Frankfurt School—have indicated, certain noun abstractions, what Adorno called the "Jargon of Authenticity," were fostered by advertisers, political propa-gandists, and even more subtly by existentialist philosophers. In Adorno's reckoning, such jargon paved the way for fascism, which in common with all demagogy exploits emotional rather than logical responses to language. Adorno identifies the key problem with abstract nouns used as catchphrases or slogans: "[t]he dialectic is broken off: the dialectic between word and thing as well as the dialectic, within language, between the individual word and their relations."[53] He describes, for instance, how existentialists charge certain words with religious sacredness to give those concepts authority where none otherwise would logically exist, so that

> [p]rior to any consideration of particular content, this language molds thought. As a consequence, that thought accommodates itself to the goal of subordination even where it aspires to resist that goal. The authority of the absolute is overthrown by absolutized authority. Fascism was not simply a conspiracy—although it was that—but it was something that came to life in the course of a powerful social development. Language provides it with a refuge. Within this refuge a smoldering evil expresses itself as through it were a salvation.[54]

The phrase "the course of a powerful social development" refers in part to existentialism, which mystifies domination "while pretending to be a cri-tique of human alienation." It does this, according to Adorno, by defining "freedom," "subjectivity," and "authenticity" without regard to the historical conditions that constitute and limit such concepts. Without a comparison to historical and social conditions, those concepts appear autonomous, with-out content, mystified. Critical theory attempts to reconstruct the historical forms of domination that existentialism and the language of commercial and political modernity obscure.

Of course, *Watt* avoids documentary analysis that might reconstruct historical forms of domination; Beckett approaches such an end negatively, by mocking Watt's existentialist interpretation of employment and his fallacious explanation of the vestiges of historical exploitation in the poverty that abounds for "miles around in every conceivable direction."[55] For example, when considering why servants work downstairs before working upstairs, Watt first reasons that the term of their employment must match the movements of the sun, but finding this absurd, he comes up with another logic. He converts employees into nouns, "short-time men" and "long-time men," "ground-floor men" and "first-floor men," and reckons that the "period and distribution of service must depend on the servant, on his abilities, and on his needs."[56] The absurdity of this construction—that understands employment only from the perspective of the employee's individual characteristics converted into a pseudo-meaningful "thing"—mocks an existentialist doctrine that makes a fetish of the personality without understanding subjectivity in relation to historical reality. These nouns are empty and misleading, because obviously the tenure of employment depends also on the employer, the availability of jobs, and a host of other historical and economic factors.

Similarly, in a fantastic satire of anthropomorphized history, Watt comically explains the squalid contemporary poverty around the big house as a function of Mr. Knott's generosity. The historical explanation emerges when Watt tries to understand how it has come about that the ground-floor servant administers Mr. Knott's dinner leftovers to a local "famished" dog. Watt reasons that in "the long distant past" Mr. Knott must have set aside an annuity to support "colony of famished dogs," and to look after them, a local catholic family, who actively pursue a combined age of 1000 through logarithmic rates of propagation. By putting charity as the first cause of economic squalor rather than mitigating decency Watt caricatures a view of the Anglo-Irish ascendancy, held by W.B. Yeats and others, that they operated mostly from a spirit of *noblesse oblige* rather than or as well as economic greed through exploitation. Thematically Beckett satirizes ideological versions of the Irish past, but formally that satire pins down the logical fallacy that Nietzsche assigned to the crude fetishism in language itself:

> Everywhere it sees a doer and doing; it believes in will as *the* cause; it believes in the ego, in the ego as being, in the ego as substance, and it projects this faith in the ego-substance upon all things—only thereby does it first *create* the concept of "thing." Everywhere "being" is projected by thought, pushed underneath, as the cause.[57]

As the narrator tells us, echoing Nietzsche directly: Watt "laboured to know [...] which the doer, and what the doer, and what the doing, and which the sufferer, and what the sufferer, and what the suffering."[58] This form of identification—either to identify a victim or "benefactor"—does very little to get at the more elusive economic relationships that create or ease suffering, in the case of Ireland, namely *lassiez faire* liberal economics, which mid-nineteenth century at least, arguably made worse the lives of more than simply "famished" dogs. The text reveals how Watt's deference to figurehead authority, to institutional agency, leads him to anthropomorphize history and irrationally interpret events.

Watt is of course the bourgeois middleman: he is a sort of Trinculo figure, easily seduced. University educated, he knows about technology, literature, and linguistics. From a certain perspective, he epitomizes the problem facing Europe in the 1930s, that the middle-class became the primary ranks from which fascism drew support.[59] And when Watt does challenge authority, breaking a rule in the house, for instance, refusing to watch over the dog's meal, we get a glimpse of how extraordinarily his behavior has become regulated:

> No punishment fell on Watt, no thunderbolt, and Mr. Knott's establishment swam on, through the unruffled nights and days, with all its customary serenity. And this was a great source of wonder, to Watt, that he had infringed, with impunity, such a venerable tradition, or institution. But he was not so foolish as to found in this a principle of conduct, or a precedent of rebelliousness, ho no, for Watt was only too willing to do as he was told, and as custom required, at all times. And when he was forced to transgress, as in the matter of witnessing the dog's meal, then he was at pains to transgress in such a way, and to surround his transgression with such precautions, such delicacies, that it was almost as if he had not transgressed at all.[60]

The institution for which Watt works has become to him so divinely mystified that he expects a "thunderbolt" for his misbehavior. Watt's assumptions depend upon a rigorous expectation that impunity will be followed by punishment, such that he stills his mind with the thought that "if he went unpunished for the moment, he would not perhaps always go unpunished."[61] Watt has failed to attend the dog's meal not for a rational reason, based upon a logic of resistance or "rebelliousness," but because he had "no love for dogs." The narrator reminds us, in other words, that a form of critique, resistance, and rebellion is available: Watt will not, however, be the one to do it. And his minor rebellion, driven by petty concerns, ignores the larger, widespread problems to do with hunger and poverty.

In the third chapter "Sam" and Watt are both interned. The biographical overtones of the narrator's name are patent. Marjorie Perloff has suggested that the strange communication between the characters—*Watt* begins to speak backwards, to invert the order of the letters, the words in the sentence—reflects Beckett's work as a courier in the resistance transmitting war code,[62] and W.J. McCormack argues that the episode suggests two prisoners communicating in Curragh camp during the Irish civil war, or recalls some kind of gulag.[63] Since Sam is aware that Watt should have understood historical aggression, should have come to a critique of word-superstition, and should have made conscious his methods of rebellion—but resisted doing all of these—to read Watt himself as a resistance prisoner makes no sense at all. When Sam writes that "we knew resistance too, resistance to the call of the kind of weather we liked,"[64] he simultaneously evokes the resistance in his language, just as the surroundings themselves evoke the punishment for resisting, yet he uses the word merely in the context of resisting one's impulse to go out in good weather. If we were to read the book biographically we might conclude that Watt is the antithetical opposite of Sam Beckett, who indeed overcame his own bourgeois apathy, his view of historical aggression as accidental; he fought fascism and devoted himself to diagnosing the linguistic mystifications of bourgeois modernity, which provides fascism a refuge. That the camp can be read as both mental institution and internment camp grimly compresses the alternatives: if one resists, one could end up in a camp; if one serves, rationalizes the mystifications of power, one participates in an irrational madness. Perhaps this was the choice Beckett felt he himself faced.

A better explanation for Watt's convoluted language in the camp could be found in Adorno's recognition that the misuse of language in philosophy, commerce, and politics provides fascism a refuge, and "within this refuge a smoldering evil expresses itself as through it were a salvation."[65] In a sense Mr. Knott becomes that salvation, just as in a different sense *Godot* does, and Watt's language becomes progressively more convoluted the closer he comes to describing Mr. Knott. The following passage, when Watt has begun to invert the "letters in the word together with that of the sentences in the period," is a case in point:

> *Lit yad mac, ot og. Ton taw, ton tonk. Ton dob, ton trips. Ton vila, ton deda. Ton kawa, ton pelsa. Ton das, don yag. Os devil, rof mit.*
> This meant nothing to me. [...]
> Thus I missed I presume much I suppose of great interest touching the [...] closing period of Watt's stay in Mr. Knott's house.
> But in the end I understood.[66]

The passage can be "translated" as follows:

> So lived for tim. Not sad, not gay. Not awak, not aslep. Not aliv, not
> aded. Not bod, Not spirit. Not Wat, not knot. Til day cam, to go.

We should recall Adorno's claim about the effects of language when governed
by the mystifications of jargon: "[t]he dialectic is broken off: the dialectic
between word and thing as well as the dialectic, within language, between
the individual word and their relations."[67] Think of the five sandwichmen
marching across the streets of Dublin in *Ulysses,* each sporting a tall white
hat with a different letter H.E.L.Y.S., to advertise Hely's stationary. Joyce
performs the effects that advertising jargon has upon language by having
the word march across the city, free from its relations in language. And as
the men become separated, Y "lagging behind," syntactical and, in this case,
orthographic relations comically scramble.[68] Similarly, Watt's words can be
made to make sense, but as he describes the mystery of this metaphysical,
political, and theological entity—not alive, not dead, not body, not spirit—
his words lose their relation within the sentence, and his letters within the
word, as if demonstrating the exaggerated effects upon language within this
institution. The dialectic between word and experience disintegrates because
there is nothing presented to the senses; Mr. Knott is naught.

From this perspective—that Beckett appears most interested in diagnosing the linguistic mystifications that lead bourgeois man to rationalize
the irrational—setting the book in Ireland, one of the few neutral countries in Europe during World War Two, makes more sense. *Watt* investi-
gates the strategies that allow one to profit from complacency, valorizing
doing nothing, defending the status quo simply because one benefits from
it, and rationalizing the past in a manner that exculpates one from addressing injustices in the present. Beckett devotes his work not to resistance, but
to the strategies by which Watt is seduced and to the rationalizations that
Watt develops not to resist. Neutrality makes an apt analogy to forgoing
individual responsibility to see through propaganda and evaluate its consequences. Simultaneously, and vexingly, Beckett's experimental style presents
the reader in the form of literary and aesthetic conundrums similar interpretative challenges to those propaganda presents: meaning recontextualized.
Should we consider, we ask, the author's biography, the material Beckett
has omitted because of "fatigue and disgust," the history of a period whose
war propaganda, ideological histories, and perceptions of economic victim-
hood eventually stretched the world on the wrack of barbarous warfare?
The answer is first yes and then no, because just as the book provokes us to
consider such material, it leaves us still struggling with the insufficiency of

the logic. And in that moment of misrecognition Beckett exposes how propaganda works, as material familiar from another and important context is reframed for a different ideological end. The difference here, of course, is that the book makes us aware of such a movement, requires us to know history and biography, even identify a compositional process that involves erasure, the better to recognize how such narrative decisions affect us well beyond the literary text.

Notes

1. These vocabulary notebooks are held in the Beckett International Foundation at the Reading University Library, UoR 5006. © Samuel Beckett Estate.
2. The six notebooks comprising the German diaries are also housed at the BIF at the Reading University Library. All citations hereafter will include notebook number and date. © Samuel Beckett Estate.
3. James Knowlson, *Damned to Fame: The Life of Samuel Beckett* (London: Bloomsbury, 1996), 297–318.
4. Deirdre Bair, *Samuel Beckett: A Biography* (New York: Harcourt Brace Jovanovich, 1978), 314.
5. Ruby Cohn, *A Beckett Canon* (Ann Arbor: University of Michigan Press, 2001), 113.
6. Marjorie Levinson, "What Is New Formalism?" *PMLA* 122.2 (2007): 558–69.
7. Richard Strier, "How Formalism Became a Dirty Word, and Why We Can't Do Without It," Afterword to *Renaissance Literature and Its Formal Engagements*, ed. Mark David Rasmussen (New York: Palgrave, 2002), 213.
8. Ibid., 212.
9. Kiernan Ryan, *New Historicism and Cultural Materialism: A Reader* (London: Arnold, 1996), xvii.
10. Near the trend's beginning, see Michael North, *The Political Aesthetic of Yeats, Eliot, and Pound* (Cambridge: Cambridge University Press, 1991); more recently, see Vincent B Sherry, *The Great War and the Language of Modernism* (New York: Oxford University Press, 2003).
11. Samuel Beckett, *Mercier and Camier* (London: Calder, 1974), 118.
12. These manuscripts are held at the Harry Ransom Center, The University of Texas at Austin.
13. Samuel Beckett, *Watt* (London: Calder, 1998), 247.
14. Slavoj Žižek, *The Sublime Object of Ideology* (London: Verso, 1989), 50.
15. Roland Barthes, "Myth Today," in *Mythologies* (New York: Hill and Wang, 1972), 114.
16. Ibid., 118.
17. Ibid., 142.
18. Beckett, *Watt,* 246.
19. See Tobin Seibers, "Kant and the Politics of Beauty," *Philosophy and Literature* 22.1 (1998): 31–50.

20. Primo Levi, *Survival in Auschwitz: The Nazi Assault on Humanity* (New York: Simon & Schuster, 1996), 168. See also Olivier Razac, *Barbed Wire: A Political History* (New York: New Press, 2002), 65–6.
21. Denis Peschanski, *La France Des Camps* (Paris: Gallimard, 2002).
22. Knowlson, *Damned to Fame,* 314.
23. Beckett, *Watt,* 154–5.
24. Ibid., 156.
25. Concentration camps, for instance, developed first in the Boer War, and, in the camps surrounding Beckett in South France, first the French held Spanish refugees fleeing the civil war, then Germans held Jews and resistance fighters, and finally Americans held German prisoners before easily erasing them from the landscape.
26. Giorgio Agamben, *Homo Sacer: Sovereign Power and Bare Life* (Stanford: Stanford University Press, 1998), 167.
27. Beckett, *German Diaries* Notebook 4, 11 January 1937. © Samuel Beckett Estate.
28. W.J. McCormack, *From Burke to Beckett: Ascendancy, Tradition and Betrayal in Literary History* (Cork: Cork University Press, 1994), 9.
29. Agamben, *Homo Sacer,* 168–80.
30. Beckett, *German Diaries* Notebook 4, 31 January 1937.
31. Friedrich Stieve, "What the World Rejected: Hitler's Peace Offers" (Washington DC: Washington Journal, 1940).
32. Beckett, *German Diaries* Notebook 4, 15 January 1937.
33. Adolf Hitler, *Mein Kampf,* trans. Ralph Manheim (New York: Houghton Mifflin Company, 1999), 1.
34. Ibid., 14.
35. Ibid., 152.
36. See "Beckett, German Fascism, and History: The Futility of Protest," in Marius Buning et al., eds., "Historicising Beckett/Issues of Performance," *Samuel Beckett Today/Aujourd'hui* 15 (Amsterdam: Rodopi, 2005), 101–16.
37. On documentary realism, see Samuel Lynn Hynes, *The Auden Generation: Literature and Politics in England in the 1930s* (New York: Viking Press, 1977).
38. Beckett, *German Diaries* Notebook 4, 2 February 1937.
39. In a letter to Mary Manning, 16 July [?] 1937. This letter is housed in the Samuel Beckett Collection at the Harry Ransom Center, University of Texas at Austin. © Samuel Beckett Estate.
40. Gershon Weiler, *Mauthner's Critique of Language* (Cambridge: Cambridge University Press, 1970), 141.
41. See Mary Bryden, *Samuel Beckett and the Idea of God* (New York: Macmillan, 1998) and David Hesla, "The Shape of Chaos: A Reading of Beckett's *Watt,*" *Critique: Studies in Modern Fiction* 6.1 (1963): 85–105.
42. Beckett, *Watt,* 49.
43. Ibid., 56.
44. Ibid., 129.
45. Ibid., 78.

46. Ibid., 74–5.
47. Ibid., 79–80.
48. Ibid., 72.
49. Ibid., 40.
50. Ibid., 42.
51. Ibid.
52. Ibid., 42–3.
53. Theodor Adorno, *The Jargon of Authenticity* (Evanston, IL: Northwestern University Press, 1973), 12.
54. Ibid., 5.
55. Beckett, *Watt,* 97.
56. Ibid., 131.
57. Friedrich Nietzsche, "Twilight of the Idols," in *The Portable Nietzsche,* ed. Walter Kaufmann (New York: Penguin Books, 1982), 483.
58. Beckett, *Watt,* 115.
59. Eric Hobsbawm, *The Age of Extremes: A History of the World, 1914–1991* (New York: Pantheon Books, 1994), 122.
60. Beckett, *Watt,* 113–4.
61. Ibid., 114.
62. Marjorie Perloff, "Witt—Watt: The Language of Resistance/The Resistance of Language," in *Wittgenstein's Ladder* (Chicago, University of Chicago Press, 1996), 137–8.
63. McCormack, *From Burke to Beckett,* 391–2.
64. Beckett, *Watt,* 149.
65. Adorno, *Jargon of Authenticity,* 5.
66. Beckett, *Watt,* 165–6.
67. Adorno, *Jargon of Authenticity,* 12.
68. James Joyce, *Ulysses* (New York: Vintage Books, 1986), 127.

CHAPTER 4

Beckett's Theatre "After Auschwitz"

Jackie Blackman

Adorno: To write poetry after Auschwitz is barbaric.
Beckett: One hasn't the right to sing anymore?

S amuel Beckett and Theodor W. Adorno met several times,[1] but never
to participate in this fictitiously arranged dialogue. In 1961, on the second of these occasions, Beckett was in Frankfurt to attend an evening
celebration given in his honor by the Suhrkamp publishing house, where he
would hear Adorno deliver a "lengthy, profound disquisition on *Endgame*
speaking of loss of meaning, decline and decay."[2] Adorno, in conversation
with Beckett at a lunch beforehand, was more interested in discussing his
idea of "the etymology and the philosophy and the meaning of the names" in
Endgame (insisting that "Hamm" derived from "Hamlet")[3] than the ethics
of Holocaust representation. Yet, within Holocaust studies and the context
of the unsayable, Beckett's play, *Endgame* (1956), and Adorno's well-worn
dictum "poetry after Auschwitz is barbaric"[4] (1949) are often positioned
together, even though there was no chronological connection between the
two. A more definite linkage (not commonly alluded to) did come later,
however, when Adorno re-visited his judgment of "no poetry" in the light of
Beckett's exemplary autonomous art.[5]

This hazy relationship coupled with an apparent lack of scholarly attention directed towards a Jewish context in Beckett's two preceding plays,
Eleutheria and *Waiting for Godot,* indicates the difficulties attached to this
controversial topic, as well as the problems of theorizing either personal or
collective memories and testimonies. Undoubtedly, it is because of Adorno's

deep admiration for *Endgame* (in particular) that Beckett is now cited within studies on Holocaust drama as some kind of canonical authority whose aesthetic of catastrophe has been "widely emulated."[6] Yet it is notable that in all of these various discussions, notwithstanding the fact that in 1969 Beckett was awarded the Nobel Prize in part for his uniquely articulated sensitivity to wartime suffering, his personal knowledge of Jewish issues is rarely referred to.[7] Such investigation has largely been limited to the fields of biography and memoir.[8] Perhaps this is because Beckett's identity as an intellectual-outsider, witness-survivor (an Irish, literary, non-Jewish member of the French Resistance who avoided capture), positioned him away from the subjective testimonies of Holocaust survivors. Elsewhere, I have provided a biographical and thematic examination of Beckett's pre-war, wartime and postwar experience of Jews, Judaism and anti-Semitism in relation to aspects of *Eleutheria* and *Waiting for Godot.*[9] This chapter now examines those cultural conditions in relation to *Endgame*. However, the purpose is not to limit the work to a "Jewish" reading, but rather to search for new pathways between history and art that recognize Beckett and his writing as part of the one reality. Arguably, acknowledging Beckett's Jewish context makes it possible to understand more fully his achievement in creating a dramatic language, which was uniquely able to define the bleakness of his time.

In the aftermath of the Holocaust, Beckett's work was seen by many to reflect the failure of the word in such a patently inhuman world. George Steiner, a Jewish critic who wrote extensively about the Holocaust, describes Beckett's theatre as "haunted" by the fact "that the living truth is no longer sayable."[10] This failure of language not only articulated a rupture in civilization but also, just as significantly, it reflected a Europe which was unable to confront the atrocious crimes that had been committed on its own doorstep. The question, then, of why or how Beckett, as witness to this untenable situation, felt obliged or even able to paint such an early picture of what Adorno termed "the misery of philosophy,"[11] becomes an important one for art. I suggest therefore, that any analysis of Beckett's early postwar writing must be deemed incomplete without at least some consideration of "Beckett-the-writer," as both "survivor"[12] and "witness"[13] of "Auschwitz."[14]

That said, in the mid- to late forties, the Holocaust[15] was a complex matter, not a definable or describable fact, or at least only definable and describable as a fact without meaning. A new language had to be found. Beckett and his contemporary critics were collectively and generally, both witnesses and survivors of "Auschwitz" and, by default, limited by their own and others mediated memory of this catastrophic event; in other words, confined to a learned vocabulary, both visual and textual. This vocabulary came via the press and film not in gruesome Technicolor but in searing black and white. In 1945 it was silent images, not words that spoke so forcibly of the Nazi atrocities. Beckett's distinctive and timely rupture of the textual echoed a moment in history when silent images and meaningless words became the currency of catastrophe. Shoshana Felman and Dori Laub have suggested that certain wartime and post-war writers (Albert Camus for example) testify to the "cataclysmic trauma" of the Holocaust, which they in turn interpret as a "radical crisis in witnessing."[16] Arguably Beckett's post-war writing can be read in a similar light, and I would challenge critics like Alain Badiou, for example, who, in order to seek a positive aesthetic content in Beckett, glance over the topic of "Auschwitz."[17]

Sartre noted in "Situation of the Writer in 1947": "The word 'Jew' formerly designated a certain type of man; perhaps French anti-Semitism had given it a slight pejorative meaning, but it was easy to brush it off. Today one fears to use it; it sounds like a threat, an insult, or a provocation."[18] Yet Beckett, in a number of his works, did not fear to use the word "Jew," or other more complex predicates emblematic of the Jew such as "Youdi"[19] in *Molloy*,[20] or "Israelite"[21] in *Malone Dies*. Moreover, it appears that even in the face of apparent contemporary limits, he felt an obligation to highlight both *passive* and *vociferous* expressions of anti-Semitism, even if obliquely— the most intriguing example being "de goûter du bébé juif"[22] from *Premier Amour* (1946) (translated by Beckett as "a dish of sucking jew" in *First Love* [1972]).[23] Later, Beckett's use of the term "Jew from Greenland" in his first play, *Eleutheria,* as well as his last minute erasure of the heavily loaded Jewish name "Lévy" from *Godot,* confirm the presence of an acute sensibility towards Jewish issues.[24] However, it is perhaps in *Endgame,* more particularly than anywhere else in his work, that Beckett gives us veiled hints and allusions that relate to the predicament of the Jews, before, during, and after the implementation of the "Final Solution." It is the *withheld specificity* that makes *Endgame* Beckett's most far-reaching and disturbing play.

Traces of "Auschwitz"

In mid-June 1945, while he was on his first visit to Dublin since the war, Beckett received the shocking news of Alfred Péron's demise at Mauthausen.[25] When Rosette Lamont interviewed Beckett in 1983 and asked him pointedly if his use of certain images in *Godot* and *The Trilogy* related to any kind of political response to Auschwitz, she reported: "Beckett did not answer me directly, but he began to talk about his close friend Alfred Péron."[26] Beckett remembered:

During the Occupation, I was a member of the resistance. We had a spy in our midst. Most of my comrades were arrested. Alfred Péron was sent to concentration camp. At the time of liberation he was still alive. He started on a trek in the direction of France. On the roads, survivors resorted to cannibalism. Peron died of exhaustion and starvation.[27]

Just before he died in 1989, Beckett told James Knowlson a similar story, again focusing on the (often suppressed) subject of cannibalism.[28] In contrast to Péron's pitiful story,[29] Beckett had survived the remaining years of the war by escaping with his companion Suzanne Deschevaux-Dumesnil to rural Roussillon in the "free zone," and, as Knowlson observes, "Beckett [...] knew that he would never have survived such an ordeal."[30] Beckett appears as one who has been spared the fate of a number of his close Jewish friends and contacts, yet whose creative impulse is, perhaps, inextricably bound to them. That these distressing personal experiences had a major impact on Beckett's subsequent fiction, poetry and drama is an opinion currently shared by many commentators, but how is one to interpret such an influence, or indeed, make it relevant to current historicizing scholarship either within or beyond Beckett studies?[31]

During a visit to Beckett in 1985, Gordon Armstrong learned that on returning to Ireland in 1945 Beckett had visited with Jack B. Yeats and asked him in what way his painting had changed over the years. "Jack replied: 'Less conscious.'" For Beckett, "the words were like lightning bolts,"[32] and may well have suggested a methodology for Beckett to follow in terms of abstracting the unspeakable. Beckett, as a contemporary "witness" to the intense human suffering of the war, one might argue, established, from 1946, a new kind of "less conscious" aesthetic which nevertheless provided a recognizable picture of wartime atrocity and trauma; one which can be seen to utilize the "less conscious" qualities of silence and/or abstraction in order to justify the problematic drive (or will) to express the "unspeakable."

In his essay, "What is an Author?" Foucault asks: "How can one define a work amid the millions of traces left by someone after his death?"[33] I want to ask that question in a slightly different context here, to establish some kind of theoretical explanation for using biographical material in relation to a historicized reading of Beckett's work. And by that I don't necessarily mean a wholehearted return to "the-man-and-his-work-criticism," but more of a re-examination of Beckett's post-war writing in the light of those oblique traces left by him, as man and author, both *inside* and *outside* of the work in question. Foucault's assertion that the point of today's writing is "a question of creating a space into which the writing subject constantly disappears"[34] allows us to look at Beckett as writing subject—that is, the disappearance of "Beckett-the-writing-subject" into "the-subject-of-his-writing"—and to interrogate his depiction of the "Jewish Question" through the use of an erased language which disappears but is never totally silent or absent: a language of "traces" left by the so-called "dead author."

This "language of traces" can also be viewed in ethical as well as aesthetic terms. Jewish artist and camp survivor, Avigdor Arikha, whom Beckett met during the genesis of *Fin de partie* (*Endgame*) in 1956,[35] told Judith Wechsler that both he and Beckett "shared the belief that there is a 'deep connection between ethics and aesthetics.'"[36] According to Peggy Phelan, Arikha's work "allows us to think of abstraction as a response to trauma, one rooted in a rejection of the obeisance often involved in mimesis [...] for both Beckett and Arikha the question of seeing was simultaneously an ethical, aesthetic, and deeply intimate one."[37] Beckett himself said: "I want to bring poetry into drama, a poetry which has been through the void,"[38] and one might extrapolate that for both Beckett and Arikha, their vivid wartime memories and images demanded release through their art. Ten years after the war, when Arikha met Beckett he was still painting semi-figurative "metaphors of persecution"[39]—with titles such as *Convoi, Cavaliers, Chassis, Tournois*. Patrick Bowles (translator of Beckett's *Molloy*) interviewed Arikha for *The Paris Review* in 1965 and asked the painter about his early artistic attempts:

A: Painting became an urgent medium.

B: Because of the visual impact of the camps?

A: I don't know. It was urgent. We saw how values can be temporal and absolutely irrelevant. But I remember more the smell of the mud. Above all the mud. Blacks and broken browns.[40]

Convoi is "a deportation scene, the mud of the experience transmuted into earth colours."[41] "Mud" for Arikha suggests the essence of wartime memory and suffering; so too perhaps for Beckett, who in 1946 tries to envision "killing" "memories" sinking "forever in the mud."[42] Certainly, *How it is* (begun in 1958[43]), an abstract work notoriously difficult for Beckett to write, evokes the mud, abject violence, and numbering system of the camps:

[...] speechless and reafflicted with speech in the dark the mud nothing to emend here there he is again last figures the inevitable number 777777 at the instant when he buries the opener in the arse of number 777778 and is rewarded by a feeble cry cut short as we have seen by the thump on skull who on being stimulated at the same instant and in the same way by number 777776 makes his own private moan which same fate something wrong there and who at the instant when clawed in the armpit by number 777776 he sings [...][44]

Anne Atik, Arikha's wife, sketched out the shared ethical, aesthetic sensibility felt between the two men in her account of Beckett reading to Arikha from the manuscript of *Fin de partie* in 1956. Beckett began to recite, according to Atik, "what is surely one of the great monologues in twentieth century drama: '*On m'a dit l'amitié, c'est ça, l'amitié…*' When he continued, '*Je me dis…*,' going from major to minor, A. burst into tears." It was, she suggests, "one of the greatest spiritual, aesthetic discoveries of A's life."[45]

The monologue translates as:

> How easy it is. They said to me, That's friendship, yes, yes, no question, you've found it. They said to me, Here's the place, stop, raise your head and look at all that beauty. That order! They said to me, Come now, you're not a brute beast, think upon these things and you'll see how all becomes clear. And simple! They said to me, What skilled attention they get, all these dying of their wounds [...] I say to myself—sometimes, Clov, you must learn to suffer better than that if you want them to weary of punishing you—one day. I say to myself—sometimes, Clov, you must be there better than that if you want them to let you go—one day.[46]

There is no obvious linkage with Clov's monologue to "Auschwitz" yet somehow, even if intangibly, the traces remain. Borrowing Arikha's words to Bowles on the difference between "image" and "art," one can characterize the difference between traditional mimesis and Beckett's innovative *tableau vivante* in *Endgame*:

> An image is something resembling something resembling something else. Not a revealer but a reminder. A painting is a revealer…not there to remind you of another subject…it is immediate. An image is not immediate. It is mediate. It imitates something which it is not.[47]

Arikha adds, "I would wish painting to be as immediate as music" (this wish was well known to be shared by Beckett in terms of his dramatic texts). Arguably it is by revealing "the wisdom of the vision"[48] (Arikha's definition of "abstraction") in an almost painterly form that Beckett achieves such immediacy in *Endgame*. In the mid-fifties both Beckett and Arikha portray a "vision" of catastrophe in their respective media of drama and painting as figurative "marks"[49] or traces rather than representational or "determined image[s]." Arikha explains his artistic approach:

> Let's say the painter begins to sing when his hand is broken….At that point he tries communicating something about experience. Not merely

supplying information…The essential thing is to sing. Art begins when it begins to sing.[50]

Earlier, in *Endgame* Beckett formulated a similar idea, if somewhat obliquely:

HAMM: Don't sing.
CLOV: One hasn't the right to sing anymore?
HAMM: No.
CLOV: Then how can it end?
HAMM: You want to end?
CLOV: I want to sing
HAMM: I can't prevent you.[51]

These passages can be set against Adorno's much quoted decree about the impossibilities of "poetry [or song] after Auschwitz." They are not unlike the response of playwright Armand Gatti, whose *Le Chant d'amour des alphabets d'Auschwitz* (1989) unambiguously refutes "Adorno's specific contention that after Auschwitz music no longer had the right to exist."[52] When Gatti had researched and read texts telling of deportees, who on their way to the gas chamber had burst into song, his response was to highlight yet another layer of complexity: "And they sing. Auschwitz becomes song. For a few seconds (a minute perhaps) Auschwitz ceased to exist."[53] Song becomes a valid way to express the unspeakable. Gatti articulates the problems of "writing" which cannot or will not evoke "Auschwitz": "If writing cannot recall the meaning of the camp, it is meaningless—it is not writing at all."[54]

Philosopher Giorgio Agamben has also contributed to this debate with the following allegation and observation: "To say that Auschwitz is 'unsayable' or 'incomprehensible' is equivalent to *euphemien,* to adoring in silence as one does with a god. Regardless of one's intentions, this contributes to its glory. We however, 'are not ashamed of staring into the unsayable'—even at the risk of that what evil knows of itself, we can so easily find in ourselves."[55] Clearly, Beckett, who could be said to adore silence, was "not ashamed of staring into the unsayable." But how does he do this? Even if "Auschwitz" as such is not mentioned in his work, the evocative traces which are left suggest a writer actively engaged in post-Auschwitz ethics. Beckett's response to "Auschwitz" was not to deny the event but to develop an appropriate approach; one that included the techniques of erasure, traces, abstraction and so forth, and could underline the presence of a *troubling* "progress" intrinsic to basic human will and natural ambition, artistic or otherwise.

In addition, Beckett's post-Auschwitz "silent" or "less conscious" aesthetic may be seen as an acknowledgment of the difficulties inherent in "survival." Ruth Kluger, a survivor of Auschwitz comes to the problem of unsayability as a contemporary writer of literary memoir. She excuses her own and other's attempts to re-present the unspeakable: "Any event you can turn into literature becomes, as it were, speakable."[56] However, unlike Beckett, writers like Agamben and Kluger who are currently working through the ethical questions of "sayability" in relation to the Holocaust have, at least, the advantage of distance. Kluger discusses the ethical problems of Holocaust representation in her genre—the memoir—and enlightens us to the problem facing other writers who also escaped the camp situation:

> Now comes the problem of this survivor story, as of all such stories: we start writing because we want to tell of the great catastrophe... Without meaning to, I find I have written an escape story, not only in the literal but in the pejorative sense of the word.
>
> So how can I keep my readers from feeling good about the obvious drift of my story away from the gas chambers and the killing fields and towards the post-war period, where prosperity beckons?... By virtue of survival, we belong with you, who weren't exposed to the genocidal danger, and we know that there is a black river between us and the true victims... I was with them when they were alive, but now we are separated. I write in their memory, and yet my account unavoidably turns into some kind of triumph of life.[57]

The triumph of the survivor is what Beckett in his postwar plays manages to avoid. His are "survival stories" which inhabit a zone light-years away from the "American dream" (where Kluger's triumphant story ends in 2000).[58] Beckett's "instantaneous myths"[59] seem culled from an infinitely less corporeal world, construed in *Ping* as: "traces blurs signs no meaning light grey almost white"[60]—"mere sensations" perhaps, but nevertheless, according to Schopenhauer, "the data from which the intuitive perceptions of knowledge comes first about in the understanding."[61]

Adorno and Beckett "After Auschwitz"

As well as finding autonomous exemplary "form" in Beckett's play, Adorno (who was of a German/Jewish background) also dealt with a number of the oblique Jewish references in *Endgame* (such as the Jewish tailor joke). However, even though Adorno's essay was unique and influential in terms of its depth, political approach, and Jewish referencing, it was not until 1966, five years after its publication in 1961, that Adorno pointedly linked *Endgame* with the Holocaust. It appears that through rethinking Beckett's work Adorno re-examined his own well-worn dictum of 1949. He stated that

> The most far-out dictum from Beckett's *Endgame,* that there really is not so much to be feared any more, reacts to a practice whose first sample was given in the concentration camps…What the sadists in the camps foretold their victims, "Tomorrow you'll be wiggling skyward as smoke from this chimney," bespeaks the indifference of each individual life that is the direction of history.

This realization led to the thought that

> Perennial suffering has as much right to expression as a tortured man has to scream; hence it may have been wrong to say that after Auschwitz you could no longer write poems.[62]

So Adorno, who once admired Beckett's work as truly autonomous, in 1966 attached *Endgame* to a very specific context—representing the valid poetic voice for the "tortured" witness/victim of Auschwitz. Historical background might be seen as important here, in that the shocking Auschwitz trials which took place in Adorno's hometown of Frankfurt may well have occasioned his change of heart. The trials started on December 20, 1963 and ran until August 20, 1965 (recorded as the longest in German legal history up to that point). They were unique in that it was the first time that the German authorities had tried their own people for committing camp atrocities, and most of the twenty defendants came from middle class, educated families and claimed they were as innocent as their victims. Lawrence Langer who was present described the proceedings as a

> …bizarre courtroom drama…[where the] scenes remain episodic and anecdotal; scenarios never coalesce; characters stay vague, as protagonists dissolve into helpless victims…while antagonists dissolve into mistaken identities or innocent puppets manoeuvred from afar.[63]

This string of adjectives could indeed partially describe the "bizarre drama" of Beckett's *Endgame.* As with the Nuremburg trials in 1946 and the Eichmann trial in 1961, the Auschwitz trials in 1963–1965 would have brought the visual and textual vocabularies of the camps back into public consciousness.

Following the trials, Adorno questions

> whether after Auschwitz you can go on living—especially whether one who escaped by accident, one who by rights should have been killed, may go on living. His mere survival calls for coldness, the basic principle of bourgeois subjectivity, without which there could have been no Auschwitz; this is the drastic guilt of him who was spared.[64]

Of course we will never know if Beckett shares Adorno's perspective in this respect. But certainly the litany of words such as *ashes, transports, rat, naked bodies, extinguished* and *exterminated,* and phrases such as *the whole place stinks of corpses,* deployed in *Endgame* places the dramaturgy firmly within an extermination camp aesthetic of horror. Dominant symbols of Nazi persecution and their sinister connotations (as described by Primo Levi) correlate with the *figurative traces* found in *Endgame:*

> [...] contempt for the fundamental equality of rights among all human beings is shown by a mass of symbolic details, starting with the Auschwitz tattoo and going all the way to the use in the gas chambers of the poison originally produced to disinfect the holds of ships invaded by rats. The sacrilegious exploitation of the corpses, and their ashes, remains the unique appendage of Hitlerian Germany...[65]

Beckett gives similar details an absurd treatment, avoiding any coherent allegory:

> *HAMM*: A rat! Are there still rats?
> *CLOV*: In the kitchen there's one.
> *HAMM*: And you haven't exterminated him?[66]

The poison (Zyclon-B) referred to by Levi was originally used in the camps to fumigate the piles of worn clothing which had been stripped from the dead. It was later used for extermination in the gas chambers, where several tins of the chemical were emptied in through the roof.[67] In *Endgame,* a play, where in the words of Arikha, "significance clashes with visual impact,"[68] Clov fumigates his own clothes with a "sprinkling tin."

> *HAMM*: A flea! Are there still fleas?
> *CLOV*: On me there's one. [*Scratching.*] Unless it's a crablouse.
> *HAMM*: [*Very perturbed.*] But humanity might start from there all over again! Catch him, for the love of God!

CLOV: I'll go and get the powder. [Exit CLOV.]
HAMM: A flea! This is awful! What a day!
CLOV: I'm back again, with the insecticide. [*Enter* CLOV *with a sprinkling tin.*]
HAMM: Let him have it![69]

Repeatedly in Beckett's early postwar writing and also later works, such as *How it is, Ping, Lessness, Fizzles,*[70] *The Lost Ones,*[71] and *Catastrophe,* our attention is drawn back to verbal or visual markers evocative of the camps.

When Beckett wrote, "I believe in progress, I believe in silence,"[72] he indicated his belief in a progress, which was not to be sought by a traditional use of language: theorizing, categorizing, quantifying, or logically narrating. He appears as a writer only too aware of how the Enlightenment pursuit of scientific progress could be appropriated to justify Fascist ideals. Yet, he was not advocating total silence either. Beckett's unique kind of silence was born out of a consciousness of the inadequacy or impossibility of rational understanding. His style of *blurring traces,* which consistently and emphatically resists appropriation, nevertheless, forces a confrontation with the core ethical challenge of the twentieth century—"that what evil knows of itself, we can so easily find in ourselves."[73] Terry Eagleton has noted of Beckett's work:

> Negativity is the way Beckett, whose art is profoundly anti-fascist without needing to speak of Nazis, maintains a secret pact with failure and finitude, of which the prime signifier is the material body...If it is an art after Auschwitz, it is because it keeps faith with silence and terror by paring itself almost to vanishing point.[74]

Beckett, realizing that "any event you can turn into literature [or drama] becomes, as it were, speakable,"[75] set about finding new dimensions within which to address humankind's plight after Auschwitz. In 1961, he came tantalizingly close to explaining his own approach to theatre ethics while addressing a group of students in Bielefeld, Germany:

> for me, the theatre is not a moral institution in Schiller's sense. I want neither to instruct nor to improve nor to keep people from getting bored. I want to bring poetry into drama, a poetry which has been through the void and makes a new start in a new room-space. I think in new dimensions and basically am not very worried about whether I can be followed. I couldn't give the answers, which were hoped for. There are no easy solutions.[76]

It was critics like Steiner and Adorno who, in the late fifties and early six-ties, recognized the power of Beckett's so-called "silence" in relation to "Auschwitz." Yet even they were circumspect in their interpretations. With hindsight, Beckett's writing can now be read as an ethical engagement with history—taciturn perhaps, but definitely not silent in the face of man's inhu-manity. By leaving traces of "Auschwitz" Beckett's early post-war drama suggests that conscious aesthetic choices have been made. His theatre, in this way, acknowledges the usefulness of a poetic language, which "has been through the void" and reaches, however inconceivably, beyond the "misery of philosophy" "after Auschwitz." When Adorno admitted, "it might have been wrong to say that after Auschwitz you could no longer write poems"[77] he had almost come to the end of his life. He would not have the time or the bio-graphical information to re-examine specifically how or why Beckett might have managed to express so clearly the perennial suffering of "tortured man." Such a reading would not only have been thought to be ultimately limiting in terms of Beckett's genius but, more importantly, it would perhaps have been seen to "ruin the unique and unsayable character of Auschwitz."[78] In 1961, George Steiner wisely said "the black fantasies of Samuel Beckett are so close to us in date as to make any judgement precarious."[79] However, I sug-gest, that more than fifty years on, it is indeed possible to read Beckett's early plays as a very personal and unique creative response to the unsayability and unplayability of the Holocaust narrative—a story impossible to represent, yet, a story which must be repeatedly told and somehow understood.

Notes

1. Stefan Muller-Doohm, trans. Rodney Livingstone, *Adorno: A Biography* (Cambridge: Polity Press, 2005), 357–58.
2. See James Knowlson, *Damned to Fame* (London: Bloomsbury, 1996), 429. Adorno's "disquisition" was subsequently published in English as "Towards an Understanding of *Endgame*," ed. Bell Gale Chevigny, trans. Samuel M. Weber, *Twentieth Century Interpretations of Endgame* (New Jersey: Prentice-Hall, 1969), 82–114.
3. Barbara Bray, a close friend of Beckett's from the late 50s, told me: "Beckett admired Adorno and his work, and talked about him often" (interview with Bray, 2007). Yet he was unimpressed with Adorno's persistence with Hamm as Hamlet and corrected him at their meeting earlier in the day (Knowlson, *Damned to Fame,* 428). Later, Adorno presented Beckett with an inscribed type-script of the lecture as "an apology [...] and as a small token of esteem." *Sotheby's: English Literature.* (Auction Catalogue) London: Sotheby's, 8 July, 2004.
4. Theodor W. Adorno, "Cultural Criticism and Society," in *Prisms,* trans. Samuel Weber and Sherry Weber (Cambridge, MA: MIT Press, 1981), 34.

5. Theodor Adorno, "Meditations on Metaphysics: After Auschwitz," in *The Adorno Reader,* ed. Brian O'Connor (Oxford: Blackwell, 2000), 86.

6. Edward Isser, *Stages of Annihilation: Theatrical Representations of the Holocaust* (New Jersey: Fairleigh Dickenson University Press, 1997), 3.

7. Here, I limit my discussion to the "Jewish" experience of the Holocaust, taking my lead from the fact that Beckett joined the French Resistance because of the ill treatment of the Jews in Paris. However, Beckett's knowledge of the war and the Holocaust would have been wider in range, and his response to suffering was all-inclusive and non-partisan. See my essay, "Postwar Beckett: Resistance, Commitment or Communist Krap?" in *Beckett and Ethics,* ed. Russell Smith (London: Continuum, forthcoming).

8. Significant scholarly exceptions in relation to *Endgame* are W.J. McCormack's "Seeing Darkly, Notes on T. W. Adorno and Samuel Beckett" in *Hermathena: A Trinity College Dublin Review* 141.2 (1986): 22–44, and *From Burke to Beckett: Ascendancy Tradition and Betrayal in Literary History* (Cork: Cork University Press, 1994). Also Anna Shildo's PhD thesis, "Tragic Elements in Samuel Beckett's Dramatic Theory and Practice," (Tel Aviv University, 1993).

9. Jackie Blackman, "Beckett Judaizing Beckett: 'A Jew from Greenland' in Paris," in Dirk Van Hulle and Mark. Nixon eds., "'All Sturm and no Drang': Beckett and Romanticism/Beckett at Reading 2006," *Samuel Beckett Today/ Aujourd'hui* 18 (Amsterdam: Rodopi, 2007), 325–40.

10. George Steiner, "Silence and the Poet," in *Language and Silence and other Essays* (New Haven: Yale University Press, 1998), 52.

11. Adorno, "Meditations on Metaphysics," 331.

12. I see Beckett as a "survivor" in terms of his narrow escape from deportation, yet also note that there will be critics who would prefer to limit the label "survivor" to those who actually spent time in a concentration camp. Lawrence Langer suggests that after the camps, "We are forced to redefine the meaning of survival, as the positive idea of staying alive is usurped by the negative one of fending off death." See "Pursuit of Death in Holocaust Language," *Partisan Review* 3 (Summer 2001): 379. Arguably, Beckett, as man and writer, realized that in the words of Geoffrey Hartman "The way we write about the Shoah [Holocaust] has a bearing on the viability of culture after the Shoah." See "Shoah and the Intellectual Witness," *Partisan Review* 1 (Winter 1998): 47.

13. John Banville has written: "If Kafka is the prophet of the 20th century, Beckett is surely its witness. His mature sensibility was formed during, and to a large extent by, the Second World War." See "Words from the Witness," *The Irish Times Weekend Review,* 25, September, 2004, 10. By 1944, according to Knowlson, when Marcel Lob was "arrested in Apt as a Jew" and "taken to the camp at Drancy near Paris, Suzanne and Beckett [would have been] under no illusions as to what his likely fate would be." Knowlson, *Damned to Fame,* 305.

14. I use the term "Auschwitz" as emblematic of the concentration camp universe.

15. For a comprehensive history of the use of the word "Holocaust," see John Petrie, "The Secular Word 'HOLOCAUST': Scholarly Sacralization, Twentieth Century Meanings," http://www.berkeleyinternet.com/holocaust/.

16. Shoshana Felman and Dori Laub, *Testimony: Crises of Witnessing in Literature, Psychoanalysis, and History* (New York: Routledge, 1992), xvii.

17. Badiou says of *How it is:* "We cannot understand the text if we immediately see it as a concentration camp [*concentrationnaire*] allegory of the dirty and diseased human animal." *On Beckett: Dissymetries,* ed. and trans. Nina Power and Alberto Toscano (Manchester: Clinamen Press, 2003), 46.

18. Jean Paul Sartre, *What Is Literature* (London: Methuen, 1967), 209.

19. Anthony Uhlmann has suggested that Beckett's use of the term "Youdi" (the offensive French slang word for "Jew") as the source of Moran's fear in the novel *Molloy* might have had an effect on "French readers who still would have had the unbelievable revelations of the Holocaust ringing in their ears." *Beckett and Poststructuralism* (Cambridge: Cambridge University Press, 1999), 48.

20. Leslie Hill writes: "it is pertinent to recall that Beckett places *Molloy* in a world that bears the marks of anti-Semitism and the diaspora," in "Samuel Beckett (1906–1989)," *Radical Philosophy—Summer 1990,* http://www.radicalphilosophy.com/default.asp?channel_id=2191&editorial_id=9833.

21. During the war, "Israelite" was the "polite" term used for the long-established French Jew who initially received protection from Vichy, in contrast to the immigrant Jews, known as "Juifs" who were the first to be rounded-up. See Carmen Callil, *Bad Faith: A Forgotten History of Family and Fatherland* (London: Jonathan Cape, 2006), 235. Beckett would have been aware of the consequences of this difference through the plight of his "Juif" friends, Paul Léon (a Russian Jew) and Henri Hayden (a converted Polish Jew). Special thanks go to Marc Caplan for pointing me towards the term "Israelite" in *Malone Dies* (London: Penguin, 1965), 54.

22. Samuel Beckett, *Premier Amour* (Paris: Les Editions de Minuit, 2003), 51.

23. See illustration of "'Jewish Pig' Wittenberg [showing two child-like naked figures sucking a pig's teats]. The association of Jews with pigs became a staple of Central-European anti-Jewish symbolism." See Steven Beller, *Antisemitism: A Very Short Introduction* (Oxford: Oxford University Press 2007), 22. Interestingly, the autograph manuscript of "Premier Amour" includes translations into French of two tributes to Alfred Péron which appeared in the *Irish Times.* See Carlton Lake, *No Symbols Where None Intended* (Austin: The Humanities Research Center, University of Texas, 1984), 147.

24. Blackman, "Beckett Judaizing Beckett," 331–35.

25. Deirdre Bair, *Samuel Beckett* (London: Vintage, 2002), 361. Beckett wrote to George Reavey on 20 June 1945 of Péron's death in Switzerland. Beckett/Reavey letters, Harry Ransom Humanities Research Center, University of Texas at Austin. © Samuel Beckett Estate.

26. Rosette C. Lamont, "Samuel Beckett's Wandering Jew," in *Reflections of the Holocaust in Art and Literature,* ed. Randolph L. Braham (Boulder: Social Science Monographs, 1990), 37.

27. Ibid., 43.

28. "At the end of the war, it was terrible! The forces just opened up the extermination camps as they came through. They had nothing to eat, those of them who were left alive. So there was cannibalism. Alfred [Péron] wouldn't do it. Amazingly he got as far as Switzerland and then he died of malnutrition and exhaustion." Knowlson, *Beckett Remembering, Remembering Beckett* (New York: Arcade, 2007), 279.

29. Georges Loustaunau-Lacau tells the story of "The unfortunate Péron, the best among us, who everybody likes, who everybody helps, who never harmed anyone. He kept going to the end and died exhausted just as France was appearing on the horizon." Loustaunau-Lacau also intimates the presence of cannibalism on the long march with Péron, "The S.S. shot one of our group who had moved two meters out of line to pick up a potato in a field. We were allowed to bury him there and then so we dug a hole deep enough to prevent the crows, the Poles and the Russians from digging him up. Had a badly buried Yugoslav not been half devoured the previous evening. *Chiens Maudits: Souvenirs d'un rescapé des bagnes hitlériens* (Paris: Editions du Réseau Alliance, 1945), 89–91. Translated for this essay by Mary Richardson.

30. Knowlson, *Damned to Fame,* 345.

31. I say "beyond" since we can now find Beckett being cited by historians. See, for example, Clair Wills, *That Neutral Island: A Cultural History of Ireland During the Second World War* (London: Faber & Faber, 2007), 413–14, and Richard Vinen, *The Unfree French: Life Under the Occupation* (London: Penguin, 2007), 367–76.

32. Gordon Scott Armstrong, *Theatre and Consciousness: The Nature of Bio-Evolutionary in the Arts* (New York: Peter Lang, 2003), 33.

33. Michel Foucault, *The Foucault Reader,* ed. Paul Rabinow (London: Penguin, 1991), 103.

34. Ibid., 102.

35. Arikha, at the age of 15, after three years in captivity, had escaped from a concentration camp in the Ukraine to Palestine under the protection of the Red Cross and through the agency of the Youth Aliyah in 1944. See Patrick Bowles, "A Portfolio and an Interview," *The Paris Review* 33 (Winter-Spring 1965): 46–52 and Avidgdor Arikha, *Boyhood Drawings Made in Deportation* (Paris: Les Amis de l'Aliya des Jeunes, 1971). While at the camp, Arihka made drawings of the "SS brutalities and massacres" that he witnessed.

36. Judith Weschler, "Illustrating Samuel Beckett," in *Art Journal* 10.23 (Winter 1993), http://findarticles.com/p/articles/mi_m0425/is_n4_v52/ai_14970137/pg_4.

37. Peggy Phelan, "Beckett and Avigdor Arikha," in *A Passion for Painting*, ed. Fionnuala Croke (Dublin: Holberton, 2006), 99.

38. Samuel Beckett quoted by Knowlson, *Damned to Fame,* 427. See note 10, 709. Beckett's words were printed in *Spectaculum* 6 (1963): 319.

39. Richard Channin, Andre Fermigier, Robert Hughes, Jane Livingston, Barbara Rose, and Samuel Beckett, *Arikha* (Paris: Herman, 1985), 208.

40. Bowles, "A Portfolio and an Interview," 46–47.
41. Mordechai Omer, *Avigdor Arikha: Drawings* (Tel Aviv: Museum of Art, 1998), 28.
42. Samuel Beckett, "The Expelled," in *First Love and Other Novellas,* ed. Gerry Dukes (London: Penguin, 2000), 32.
43. John Pilling, *A Samuel Beckett Chronology* (London: Palgrave, 2006), 143.
44. Samuel Beckett, *How it is* (London: Calder, 1996), 153.
45. Anne Atik, *How It Was: A Memoir of Samuel Beckett* (London: Faber & Faber, 2001), 41.
46. Beckett, *Endgame,* 51.
47. Bowles, "A Portfolio and an Interview," 50–51.
48. Ibid., 47.
49. Arikha told me that rather than using a picture on the wall for his set design of *Endgame* (New York, 1984), he left only "the mark of the frame" (interview with Arikha, October 2003).
50. Bowles, "A Portfolio and an Interview," 49.
51. Beckett, *Endgame,* 46.
52. Dorothy Knowles, "Armand Gatti and the silence of the 1059 days of Auschwitz," in *Staging the Holocaust,* ed. Claude Schumacher (Cambridge: Cambridge University Press, 1998), 213. Gatti himself was a Resistance member and camp survivor.
53. Ibid., 215.
54. Ibid.
55. Giorgio Agamben, *Remnants of Auschwitz: The Witness and the Archive* (New York: Zone Books, 1999), 32–33.
56 Ruth Kluger, *Landscapes of Memory: A Holocaust Girlhood Remembered* (London: Bloomsbury, 2003), 141.
57. Ibid., 164–65.
58. When Kluger's position was Professor Emerita at the University of California, Irvine.
59. Barbara Bray's term for both *Waiting for Godot* and *Endgame.* Interview with the author, Sept 2007.
60. *Ping,* in *Samuel Beckett: Poems, Short Fiction, Criticism* (New York: Grove Press, 2006), 371.
61. Arthur Schopenhauer, *On Vision and Colors: An Essay,* trans. by E.F.J. Payne, ed. David E. Cartwright (Oxford: Berg, 1994), 21.
62. Adorno, " Meditations on Metaphysics," 86.
63. Lawrence. L. Langer, *Admitting the Holocaust: Collected Essays* (New York: Oxford University Press, 1975), 89.
64. *The Adorno Reader,* ed. Brian O'Connor (Oxford: Blackwell, 2000), 86–87.
65. Primo Levi, *The Mirror Maker* (London: Abacus, 2002), 165.
66. Beckett, *Endgame,* 37.
67. Walter Laqueur, ed., *The Holocaust Encyclopedia* (Yale: Yale University Press, 2001), 235.

68. Avigdor Arikha used this vivid phrase when discussing his set designs for *Endgame* (Arikha, October 2003).

69. Beckett, *Endgame,* 27.

70. In collaboration with Beckett, Arikha illustrated sections from both *Fizzles* and *The Lost Ones.*

71. This piece of fiction was dramatized by George Tabori as *Der Verwaiser* in 1981. According to Anat Feinberg, *The Lost Ones* "evoked in Tabori the idea of mass murder, the image of a man who systematically depopulates a given place. His primary association was Auschwitz." *Embodied Memory: The Theatre of George Tabori* (Iowa: University of Iowa, 1999), 137.

72. Beckett, *The Unnamable,* 124.

73. Agamben, *Remnants of Auschwitz,* 32–33.

74. Terry Eagleton, "Beckett and Nothing," in *Reflections on Beckett: A Centenary Tribute,* ed. Anna McMullan and Steve Wilmer (Ann Arbor: University of Michigan Press, forthcoming 2009).

75. Kluger, *Landscapes of Memory,* 141.

76. Beckett quoted by Knowlson, *Damned to Fame,* 427. Beckett's words were printed in *Spectaculum* l.6 (1963): 319.

77. Adorno, "Meditations on Metaphysics," 86.

78. Giorgio Agamben in *Remnants of Auschwitz* reports: "Several years ago, when I published an article on the concentration camps in a French newspaper, someone wrote a letter to the editor in which, among other crimes I was accused of having sought to 'ruin the unique and unsayable character of Auschwitz.'" *Remnants of Auschwitz,* 31. In France, even in the 1990s, Auschwitz was still considered by some to be "unsayable."

79. George Steiner, *The Death of Tragedy* (London: Faber and Faber, 1978), 349.

CHAPTER 5

"Faintly Struggling Things": Trauma, Testimony, and Inscrutable Life in Beckett's *The Unnamable*

Alysia E. Garrison

> Little by little the haze formed again, and the sense of absence, and
> the captive things began to murmur again, each one to itself, and it
> was as if nothing had ever happened or would ever happen again.
> Samuel Beckett, *Malone Dies*

As the historicist turn in Beckett studies attests, it is not enough to suggest, as Slavoj Žižek does in *Looking Awry,* that *Waiting for Godot* is simply a dramatization of nothingness and absence.[1] To do so would vacate historical reference and shut the door on the possibility of reading the text's situated allusions to a damaged time. Poststructuralism has often sublimated historical trauma as simply an instantiation of transhistorical trauma: for Žižek, the condition of trauma is equated with the Lacanian Real which then becomes the grounding for an ethics based on psychoanalysis, and for Jean-François Lyotard, trauma is linked to the sublime as the collective condition of postmodern aesthetics.[2] To redress this dangerous collapse that threatens to blur all distinctions and usher in a universalism where traumatic representation is voided of specificity and read as a universal condition to which we are all subject, recent work in trauma theory has usefully distinguished the separate concepts of structural absence and historical loss while noting their necessary tie.[3] Intellectual historian

Dominick LaCapra argues that theories of trauma have too often conflated absence and loss, conditions that circulate as undifferentiated kernels of negativity that make it difficult to parse narrative representations of historically situated trauma while recognizing aporia as the condition of possibility for such representations. The task, as set by LaCapra, is to read in posttraumatic narrative the singularity of a limit event such as the Holocaust while also opening up its conditions to rigorous analytical and comparative critique: to note the specificity of events in order to unbraid historical trauma from transhistorical trauma while addressing how these registers are tropological intimates. This essay reads such intimacy between figures of trauma in *The Unnamable,* a text that does not "name" its referent, as studies of poststructural Beckett have long suggested, but nonetheless "nods" to it in metonymic gesture. For Beckett's *oeuvre* the representation of history is a problem that will "go on" while the imprint of memory still speaks.

Trauma theory's substrate is literary expression under pressure: writing that bears the affective trace of an inscription, scar, or wound as the mark of a lost fold in history shot through with the violence of its happening, yet unable to be mourned as such. The traumatic dissociation between cognition and affect produces a scission of understanding and feeling; such cognitive failure to assimilate the senselessness of trauma leaves behind "remnants" or "traces" forming representation as marks of unlived experience. As theorists such as Cathy Caruth suggest, the experience of trauma produces a rupture or shattering break in history characterized by repetition, opacity and belatedness: in something approximating a Beckettian aporia, *it* returns, but *what it is* that returns remains ungraspable, and it always returns *too late,* disrupting the coherence of present and past. Samuel Beckett's postwar work, which limns the affective and philosophical contours around disintegrating subject-object relations, emerges from an equally incommensurable set of coordinates. Between what Adorno describes as texts marked by the "loss of the object," and "its correlative, the impoverishment of the subject," the postwar work narrates the serially returning inability to see, to say, or to know tied to the mulish mandate that one must go on "failing better" at seeing, saying, or knowing.[4] In perhaps the most lucid statement of his aesthetic written during the composition of *The Trilogy,* Beckett argues that the very inability to perceive or grasp the object of reference provides an opening that is the condition of possibility for the work of art.[5] Trauma theory grapples with a similar paradox: it is only by confronting the "object" of trauma—the aporia in history that *defies,* yet *demands* our witness—that one can begin to work through the repetition-compulsion of unassimilated experiential remainders in giving voice to the past. Because trauma cannot be grasped as discrete experience, LaCapra writes, post-traumatic art may be "ambivalent or undecidable," written in an "oblique fashion" that expresses an "opacity" that does not involve "referentiality with respect to particular historical events or figures."[6] LaCapra also notes that works by "Kafka, Celan, Beckett, or Blanchot often seem to engage in this sort of indirection or veiled allusiveness."[7]

Following LaCapra, I will argue that Beckett's work brings us into proximity with the analytic and affective fallout of trauma between transhistorical structure and historical experience. While there is no shortage of studies on post-structural Beckett, and while several critics have intimated that World War II, Beckett's Resistance work, Nazi terror, and the death camps might figure conceptually in Beckett's later aesthetic, a sustained inquiry on Beckett, trauma and testimony, situated between language and history, has not yet been undertaken.[8] Through a focused reading of *The Unnamable*, I want to develop LaCapra's implication that Samuel Beckett's writing "of terrorized disempowerment," what LaCapra calls "testimonial art" or "post-traumatic writing," bears witness to trauma not personally experienced, but that is "transmitted from intimates, or sensed in one's larger social and cultural setting," and that may come "as close as possible to the experience of abject or traumatized victims without presuming to be identical to it."[9] In short, I will suggest that *The Unnamable* provides a "feel" for situated traumatic experience—from an outside, second-hand perspective—through experimental narrative form, supplementing the factualism of testimony without presupposing a relation to it. In *The Unnamable,* the condition of absolute silence is nonetheless pierced by the intermittent acoustic pulse of a voice that demands, yet defies, our witness: it is "broken, faint, unintelligible," and yet "murmurous with muted lamentation, panting and exhaling of impossible sorrow [...] as of one buried before his time."[10] These laments supplement other traumatic embodiments that animate the familiar Beckett postwar landscape: master and slave relationships, beatings and submission at the hands of someone or something, liturgical voices and murmurs mysteriously emanating from without and within, entombed immobility or, alternately, nomad-like directionless wandering.

Such embodied "testimonial art" in Beckett's postwar narrative produces an experimental emulation of traumatic experience that for Beckett is always only the ruins of an allusion. In bearing witness to the detailed opacity of trauma, *The Unnamable* stages the inability, yet mandate, to aestheticize a form of life that I call "inscrutable life"—life both structurally absent and historically lost, the site of a rapidly shuttering dialectic without closure between historical experience and the structural impossibility of capturing this experience in language. Inscrutable, or secret life, is unread life that nonetheless sounds through silence. It is faintly audible in "the little murmur of unconsenting man, to murmur what it is their humanity stifles, the little gasp of the condemned to life, rotting in his dungeon gar-roted and racked."[11] Positioned in an aporetic relationship to knowing and being, it is life "within and motionless," the space between the referent—the thing itself—and its melancholic textual figuration.[12] It is detectable only in traces: formalizations of the broken allusion, signifier, or image that hint at what Beckett in the *Whoroscope* notebook called those vast "unseen vicis-situdes of matter."[13] It might be described as the photo negative of the figure that "hides its face," its nothingness covered by the "veil" of language, a fig-ure situated historically while also secreted away, shut off from view.[14]

Because the condition of trauma presumes an original historical event or circumstance (along with a subject who suffers and is later haunted by it), and because theories of history, subjectivity, and memory are put under var-ious levels of erasure in Beckett's subtractive aesthetics, theories of trauma or mourning might seem to lack a toehold. Yet this slippage opens a critically productive window to Beckett's art. *The Unnamable* formalizes wounded forms of life by an aesthetic of diminishment—in Adorno's words, life that resembles "flies twitching after the fly swatter has half-squashed them"—which vexes the sovereign wholeness of the subject and instantiates subjec-tivity as a linguistic and historical category produced by the narrative and biopolitical contours of history.[15] The text does this even while pushing us to read the shattering of narrative, history, and subjectivity by historically contingent traumatic experience. Reading Beckett's postwar work as an effect of the precise, detailed incomprehensibility of a crisis event—and this event's embodied inscriptions—challenges nihilist or ahistorical readings of Beckett that culminate in assessments of meaninglessness or undifferentiated suffering and lack a proper account of history and its relationship to textual memory. *The Unnamable*'s exigency to testify in the face of uncertainty—as a voice, a murmur, or semantically laden suggestion where perception and archival factualism falls away—leads us to a very different sense of the void, a grey-black that is "first murky, then frankly opaque, [but] luminous none the less."[16] *The Unnamable* struggles to hear the voice of alterity reduced to murmurs and silence that hails from "kingdom unknown," and to detect the eye that peers out and "demands sympathy, solicits attention, implores assistance."[17] In the logic of trauma theory, these are synecdoches for the figure(s) that *demand* yet *defy* our witness. Lacking the fullness of represen-tation, they allude to historically situated trauma even while performing its representational aporias, its kingdoms unknown.

Two registers of trauma demand, yet defy articulation in *The Unnamable:* the transhistorical and the historical. The narrative stages a series of trans-historical figures abandoned by the trappings of language as their included, yet excluded remainder, traumatized by the structural inability to express being outside them. Voices attempt to fill the grey-black void of existence with discourses that surround and asphyxiate the narrator(s), but they fail to help "him" articulate what is ultimately "unnamable." The narrative also formalizes historical trauma by staging broken allusions as distant metaphors for life's distorted alterity under contingent biopolitical duress. The novel does not take Nazi atrocities during the Shoah as its manifest, literal subject, addressing them directly. Rather, they surface as an oblique, alien, and disquieting subtext in *The Unnamable,* a novel which functions as an *effect* of these atrocities. As Adorno has suggested, Beckett's art leaves nothing but the uncanny residue of history, a "shabby, damaged world of images," suggestive of "the situation of the concentration camps" that Beckett "never calls by name."[18] Beckett's repudiation of allegory, combined with an increasingly minimalist aesthetic, culminates in the vacating of reference and the dissolution of relation that for Beckett is the expression of the political as such. In this domain of pure mediality, Beckett's aesthetic of nonrelation witnesses the very oscillation between *constitutive absence* (the ontological, or trauma of being) and *historical loss* (historical trauma) that testifies to historically specific trauma while challenging the suppositions of humanism, ontology, and epistemology. This narrative paradox stages an aporia—the formalization of the loss of *what was already absent* in representation. *The Unnamable* avoids the conflation of loss and absence by holding these two in subtle analytical distinction, accomplishing the deconstruction and historicism of narrative at once.

History's broken deposits then accrete on uncanny images, ghostly forms of life that are doubled. Beckett's diminished and disembodied bodies are at once *the concept of being* subjected to trauma (what I call the condition of "ontological injury"), and also *particular beings' trauma*—the trauma of *social figures,* those sufferers and casualties of the violence of Nazi atrocity. The novel testifies to the absence that is constitutive of subjectivity, and to the "lost," what Primo Levi calls "the drowned" in the concentration camp. Read as a post-traumatic narrative, the very impossibility to represent life at extreme historical limits gives rise to the possibility of testimony: testimony arises in place of either surrender to the unsayable, or the desire for complete understanding.

Transhistorical Trauma: The Impoverishment of Body and Language

In a very foundational sense, *The Unnamable* bears witness to the trauma imparted by the experience of living itself: Schopenhauer's dreary conception of life's "unspeakable pain, the wail of humanity, the triumph of evil."[19] Schopenhauer's work provided a justification for Beckett's conviction that to suffer is the norm, and in a letter to MacGreevy, Beckett comments, "Schopenhauer says 'defunctus' is a beautiful word—as long as one does not suicide."[20] Beckett's philosophy of life as punishment resonates with Schopenhauer's doctrine of suffering, which defines the "pensum" as the task of living "to be worked off."[21] Likewise, Beckett's *Proust* ends with the sentiment that there is an "invisible reality that damns the life of the body on earth as a pensum and reveals the meaning of the word: defunctus."[22] Before the Unnamable narrator can be free, he has a "pensum to discharge," a series of lessons to complete and punishments to suffer at the hands of usurpers, his "tormentors."[23] The pensum is the mandate to express in general and to speak the self in particular, executed through a series of minatory demands delivered by the Masters (alternately Mahood and Basil) and by the voices that beset the narrator. Yet *The Unnamable* likens the pensum of speaking the self through language to one without hands who has the duty to clap, or one without feet who has the duty to dance.[24] Attempts to ferret out the self by the "voices" are wrung with a violence frequently characterized as torture or brutality inflicted upon the narrator(s): "[...] these voices are not mine, nor these thoughts, but the voices and thoughts of the devils who beset me,"[25] or, "I expiate vilely, like a pig, dumb, uncomprehending, possessed of no utterance but theirs. They'll clap me in a dungeon."[26] All efforts to elucidate the self or reflect on the object are doomed to fail, for "I'll know that no matter what I say the result is the same, that I'll never be silent, never at peace."[27] The experience of trauma in the text is narrated through the violent attempt to *name,* pin down, and inflict words upon the void, paired with the inability to express the nothing that is.

Related to Beckett's portrait of life as suffering, Dominick LaCapra theorizes the condition of *transhistorical,* or *structural* trauma in analytical distinction from historical trauma. Unlike discrete historical events of trauma that took place in one's past, transhistorical trauma is a foundational experience of "suffering, disturbance, loss, or even ecstasy" that is "an unsettling condition of possibility" better conceptualized in terms of *absence* rather than of *loss.*[28] Here, pangs of suffering that may be misrecognized as loss—such as "the putative modern loss of authentic memory, true community, or oneness with Being"—are better conceived as foundational voids. Such fictions as "paradise, authentic memory, or Being" are "fictions of anteriority" in the sense that they never had metaphysical presence to begin with, resulting in the common feeling that "one has lost what one never really had."[29] In *The Unnamable,* with its attitude of "complete disintegration," these absences adhere to, and are constitutive of, ontological and epistemological systems, as studies of poststructural and nihilist Beckett have suggested.[30] Missing wholeness or completion haunts the unnamed narrator's experiences of being and knowing; he will proceed in the void only "by aporia pure and simple."[31] The text registers the inability to be and to know (a condition of impotence and ignorance) as endemic to the human; what I am calling the "trauma of being" is an always already condition of "ontological injury." Paradoxically, this traumatic aporia is the generative foundation for Beckett's art, an art that takes for granted, as the conversation playfully suggests in "Three Dialogues," "the expression that there is nothing to express, nothing with which to express, nothing from which to express, no power to express [...] together with the obligation to express."[32]

In the 1937 letter to Axel Kaun, Beckett suggests that an art of undoing language opens critical possibilities unavailable to reified forms. By dissolving the word's surface, "like [...] the sound surface, torn by enormous pauses [...] we can perceive nothing but a path of sounds suspended in giddy heights, linking unfathomable abysses of silence."[33] Beckett's aesthetic approach in *The Unnamable* proceeds through consciousness about traumatic diminishment figured through language and the body. Through a mimetically subtractive language, it enacts language's profanation, Giorgio Agamben's sense of a "language that has emancipated itself from its communicative ends and thus makes itself available for a new use," freeing means from ends by rendering language's old use inoperative.[34] In opening the space of pure mediality between means and ends—the traumatic core of language's nothing, its zero point of enunciation—Beckett's narrative provides an opening for a hint of a prayer-like silence: "In this dissonance between the means and their use," Beckett writes, "it will perhaps become possible to feel a whisper of that final music or silence that underlies All."[35] So too does it seek to rend the carapace of man and plumb the nothingness beneath it.

As the veil of language is pulled away, so is the "carnal envelope," the stiff form into which matter is pummeled.[36] Like words in a state of petrification, bodies too are mechanical puppets; the unnamable narrator will dispense with both by scattering them to the winds.[37] While being tied to a body itself imparts punishment, its subtraction reveals nothing but traumatic absence, unrecognizable formlessness in place of form. This process begins very early in *Three Novels*. In the last half of *Molloy*, Moran, failing to track Molloy, his epistemology exhausted, finds the same inscrutability in himself. As structures topple around him and his Being ruptures and disintegrates, ("a crumbling, a frenzied collapsing, of all that had always protected me from all I was always condemned to be"), Moran watches the surfacing of a hollow face from a placid lake, suggestive of both crypt and womb:

And then I saw a little globe swaying up slowly from the depths, through the quiet water, smooth at first, and scarcely paler than its escorting ripples, then little by little a face, with holes for the eyes and mouth and other wounds, and nothing to show if it was a man's face or a woman's face, a young face or an old face, or if its calm too was not an effect of the water trembling between it and the light.[38]

Beckett's citation of globe imagery may originate from Schopenhauer's *The World as Will and Idea*. Schopenhauer describes the moment of unsettling reflection of one's own ghostliness, the absence of self-presence, as such: "we find ourselves like the hollow glass globe, from out of which a voice speaks whose cause is not to be found in it, and whereas we desired to comprehend ourselves, we find, with a shudder, nothing but a vanishing spectre."[39] It is notable that the evanescent, deathly, yet peaceful face is nonetheless marked by trauma: a "poor figure," it is pockmarked with holes for eyes and mouth, described as "wounds," on an empty and moribund tableau unrecognizable of gender or age. The rising face is at once every being and no being, a ghostly layer of the onion and the onion's traumatic core. It is a proleptic image, for it will return later as the Unnamable's face. Moran's attitude toward the dispossession of self is one of resignation, an abnegation of the Will toward the suffering of Being.

This round, spherical imagery of globes, eyes, and holes is of course suggestive of castration, of loss and fragmentation, as bits and pieces of a face rise up to perform the inability of the body to cohere. The wounds for eyes allude to blindness, the concealment of matter as it gets pawed and pummeled into a subject.[40] At the same time, the surfacing face suggests sight, revelation from the depths of the unknown. What is revealed to Moran at this moment is the terrifying consciousness of his formlessness: the zero point of the dwindling bodily form. Jacques Lacan, commenting on a dream of Freud, is useful in parsing Moran's uncanny moment:

> There is a horrendous discovery here, that the flesh one never sees, the foundation of things, the other side of the head, of the face, the secretory glands par excellence, the flesh from which everything exudes, at the very heart of the mystery, the flesh in as much as it is suffering, is formless, in as much as its form in itself is something which provokes anxiety. Spectre of anxiety, identification of anxiety, the final revelation of you are this—You are this, which is so far from you, this which is the ultimate formlessness.[41]

It is notable that Lacan mentions the secretory glands as the nodal point at which one anxiously glimpses the mystery of formlessness. One thinks of the many excretions and wounds throughout *Three Novels,* as abject

Molloy "oozes" urine and sweat,[42] Mahood/Worm is covered in "pustules," and the Unnamable's tears stream from eye sockets like "liquefied brain."[43] Even when the Unnamable is able to stay quiet, he is "dribbling" from his mouth in the corner.[44] Formless, the body is perforated, porous, and leaking through its "carnal envelope." These breaches not only disintegrate the body's integrity but meld the body with the field of formless others.

The hollow globe scene is an index of the increasingly minimal body that begins with Molloy and is stripped to its barest form in *The Unnamable*. In the *Three Novels,* the physical dismantling of the body and the attenuation of language mimetically produces the analytic dismantling of subject. Life throughout the *Three Novels* gets successively more bare: Molloy's failing body is helped along by crutches, Moran's declining body forces him to crawl through the forest, Malone is impotently confined to a bed, unable even to crawl, and the composite narrator of *The Unnamable* is a patchwork of voices without and within lacking even a pronoun to anchor them. Life's subtraction happens through very specific afflictions to the body, and its diminishment brings the characters a kind of serenity. But this serenity is only partially curative, wracked as it is with suffering and representational aporia.[45] As Beckett's "gallery of moribunds" diminish in corporeal form, they also begin to meld and interpenetrate with one another in zones of indistinction,[46] suggesting that in Beckett's *Three Novels,* the subtractive body leads to a sense of perforated self, yielding a porous boundary between it and other forms of life.

With the collapse of the subject and its increasing porosity comes the analytic dismantling of the object. The "anxiety of the relation," and "all that it blinds to," reaches a fever pitch as the narrator is confronted on all sides with sheer cliffs of absence and uncertainty. The narrative famously begins with three questions that immediately disorient place, person, and time and signal the text's deferral: "Where now? Who now? When now?"[47] This vertigo of displacement registers a climate of omnidolent suffering, the incantatory keening of those in a purgatorial limbo:[48] "what should I do, in my situation, how proceed? By aporia pure and simple? Or by affirmations and negations invalidated as uttered, or sooner or later?" These yesses and noes get "shit on" in *Three Novels'* aesthetics of impoverishment, as knowledge, along with life itself, gets successively barer.[49] At the terrifying core of absence, diminished life without relationality and without the ability to narrate it, the trauma of being, *The Unnamable* calls attention to the thing not witnessed, the luminous trace that endures nonetheless, a flash of vision in blindness, *inscrutable* life. Memory, the narrator suggests, will speak in the form of a "cry" or flash up as a "collision," a rupture where a relation falters.

This sense of memory dispenses with the fiction of anteriority, the sense that we can recover a past whose contours were at some point accessible to us. Rereading memory as a site of failure, the record of a past never fully present to itself, the narrative focalizes memory's heterogeneous disjunctures that open to a field of difference, the record of what *does not* get archived. The Unnamable concedes that while "I am getting to know [the inhuman creatures]," "I do not know them all. A man may die at the age of seventy without ever having had the possibility of seeing Halley's comet."[50] "[D]istant testimony"—the cry, tear, cinder, or wound—adheres to the site where memory gasps.[51] Things are not lost to the self-mastery one never possessed: Beckett's depletion of such fictions is the condition of possibility for testimony. As the narrator suggests, "what I best see I see ill."[52] Like an onslaught of tears, the wounded and increasingly porous figurations of the body, language, and mind provide an analytic opening for the philosophy of ontology while also suggesting its inscrutable historical double.

Historical Trauma, Agamben, and the Witness

As I've suggested, *The Unnamable* testifies not only to transhistorical trauma but also to historical trauma, held in subtle distinction from each other in an interplay of forces at the same time they are entwined. More specifically, I have argued that *The Unnamable* inscribes the memory of historical loss already besieged by absence, life diminished by its own traumatic lack yet also suffering in a historically specific manner. This ethical aesthetic proceeds through negative imagery, and the sound of "a feeble murmur seeming to apologize for not being dead."[53] At the beginning of the novel, the narrator begins by conceding he knows nothing, except one thing: "I know my eyes are open, because of the tears that pour from them unceasingly."[54] He possesses awareness of his body's stiff posture only by noting its liquid coating, the tears that bathe it flowing from face to neck to sides and down the back.[55] His head betrays an uncanny resemblance to the globe surfacing from the lake that Moran witnesses: it is described as "a great smooth ball [...] featureless, but for the eyes, of which only the sockets remain," with a consistency like "mucilage."[56] Amid the castrated, porous body, the site of the breakdown of subject as object, the tears are a trace amid the grey black of the nothing that is. Markedly, tears are a *translucent* trace, the remnant not of presence but of absence, of something to come "in another present, even though it be not yet mine."[57] Tears connect to other metonyms for what remains, although these—soot, ashes, and embers—are less an absent luminosity "to come" than the charred ruins of what was. The narrator describes his hair, along with his sex and nose, as bits that have fallen "so deep that I heard nothing," "like soot still."[58] On the next page, he detects "something sighing off and on and the distant gleams of pity's fires biding their hour to promote us to ashes."[59] His tormentors watch him "like a face in the embers which they know is doomed to crumble."[60] The metonymical association between tears, soot, ashes and embers does not just link the diaphanous to the leaden, but connects both to the processes of memory, a "memory so bad, the words don't come," bearing witness to the absent presence of an unnamable, but palpable referent.[61] In an essay on mourning, Jacques Derrida speaks of the "blurred and transparent testimony borne by [a] tear."[62] So, too do these metonyms testify to a site of transhistorical and historical trauma: one of absence, and, not separate from this, of the Shoah.

While perhaps dangerous to think too readily, it is not difficult to imagine the gas chambers and the process of *selektion* by which prisoners were eliminated when reading these passages. Words in the text such as "tears," "ash," "cinders," "fire," "extermination," "Jude," "massacre," and "drowned" could be read as placeholders for specified loss, the ruins of an allusion capsized in traumatic memory. These words combine with suggestive images to produce a damaged portrait of the time in which Beckett lived: where he escaped when others were captured, and the shame that inheres in survival. James Knowlson's biography and the interviews and fragments in the James and Elizabeth Knowlson's *Beckett Remembering/Remembering Beckett* report how Beckett was acutely aware of, horrified by, and involved in combating the rise of Nazism in Germany and France in the 1930s and 40s.[63] At his friend Alfred Péron's encouragement, Beckett joined the Resistance movement as a translator in September 1941, just after his close friend Paul Léon was arrested and interned near Paris, and starved and tortured by the Germans.[64] Beckett remembers that he entered the "Gloria" group "when they were rounding up all the Jews, including their children, and gathering them in the Parc des Princes ready to send them off to extermination camps."[65] For three years during these and other Nazi atrocities of World War II, Beckett lived in the village of Roussillon in Vichy France among Jews in hiding, taking refuge from the Gestapo who were arresting members of his Resistance cell. The cell, "Gloria SMH," was betrayed by an informer, and Beckett and his wife Suzanne were forced to flee; many of his co-workers were arrested and deported to Ravensbrück and Mauthausen, but Beckett was notified and escaped just in time.[66] Had he been captured, Beckett would have also ended up in Mauthausen, the concentration camp at which Alfred Péron died.[67] Beckett took refuge in a series of safe houses including Marcel Duchamp's before finally landing in Roussillon, where he likely suffered mental strain and deep depression after coming through "an extremely traumatic series of events," worried "for money, food and cloth-ing" as "virtual prisoners" in the small town.[68] While in Roussillon, Beckett reports that he got involved with the local Resistance in the last few weeks of the war: "I remember going out at night and lying in ambush with my gun."[69] And directly after the war, Beckett witnessed actual victims, starv-ing, discharged from the concentration camps. He remembers,

> At the end of the war, it was terrible! The forces just opened up the exter-mination camps as they came through. They had nothing to eat, those of them who were left alive. So there was cannibalism. Alfred [Péron] wouldn't do it. Amazingly he got as far as Switzerland and then he died of malnutrition and exhaustion. After the war we saw quite a bit of Mania, Alfred's widow.

While not a "survivor" in the sense of the term used in Holocaust studies, Beckett nonetheless fought in the Resistance, survived close friends' and colleagues' deaths, hid with Jews, witnessed emaciated camp victims, and comforted Péron's widow after the war. While this chapter is not interested in assessing Beckett's war-time mental state, LaCapra uses the term "sec-ondary traumatization" to talk about this type of removed, yet acute experi-ence of trauma, a condition which may well have plagued Beckett.[70] Given Beckett's refusal of allegory, experience and memories would not figure "referentially" in his writing in the way they would figure in, say, an autobi-ography. Nevertheless, Beckett does not eliminate the possibility that expe-rience is "used" in writing. Knowlson reports Beckett saying that although his work was not a record of experience he did of course use it, and goes on to explain how historical fragments around the time of the Shoah, includ-ing the name of Beckett's hiding place, Roussillon, mark *Waiting for Godot,* written around the same time as *The Trilogy.*[71] Likewise, *The Unnamable* contains allusions to spying and to messengers that resonate with Beckett's Resistance work,[72] while it is also scarred with phrases such as "piled up in heaps."[73] As Worm, and then Mahood are held captive in a jar adjacent to the dreadful sounds and smells of a cattle slaughterhouse, one might think of Beckett hiding like a prisoner in Roussillon, forced to watch in the dark but not act while the nearby camps carried out one of the most profound tragedies in human history. A Resistance agent forced underground, the covert insurrectionary was now imprisoned, like some "tenth-rate Touissant L'Ouverture," the leader of the Haitian Revolution who languished and died in the French jail Fort de Joux; "Most unhappy Man of Men," Beckett had scrawled in his *Dream* notebook two decades earlier, echoing Wordsworth.[74] Along with allusions to the Holocaust and the faintest suggestion of African slavery in Toussaint's appearance coupled with Hegel's master-slave dialectic, *The Unnamable* even recollects The Trail of Tears, the forced migration of Native Americans that lead to genocide, as figures walk "Indian file."[75] What I want to suggest here is that although *The Unnamable* is not explicitly *about* the Shoah, or Beckett's experiences during the war, the novel could be read as an *effect* or a *symptom* of atrocity's conditions: as testimony that radically defies the limits of "proof," the text is marked by the annihilation of expression and the return of its ghostly remains, a testimonial form most appropriate, perhaps, to witness what remains of the human in the inhuman in the aftermath of traumatic limit events such as genocide, slavery, or famine. Tattered flashes of the Holocaust become concentrated images of the wreckages of human history: Walter Benjamin uses the term "dialectical images" to refer to the rising up of constellations that present the historical object within a force field of past and present. As *The Unnamable* reasons, time does not "pass from you" but rather accumulates: it "piles up all about you," mixing "the time of the ancient dead" with "the dead yet unborn," a heap of disaster piling toward the sky, to the text's silent witness, like Benjamin's Angel of History who watched mouth agape.[76] Part of the text's anxiety surrounding representation has to do with just these kinds of exclusions, "all that it blinds to," against the ethical imperative to testify on behalf of those who cannot do so. *The Unnamable* foregrounds how the incongruence between "blank words" and image brings about blind spots:

> [...] these nameless images I have, these imageless names, [...] how can they be represented, a life, how could that be made clear to me, here, in the dark, I call that the dark, perhaps it's azure, blank words, but I use them [...][77]

Despite Beckett's rhetoric of failure, *The Unnamable*, at least in part, succeeds in capturing what narrative "blinds to" by calling attention to the limits of expression in the text, these "nameless images" and "imageless names" that function as keys to language's silences, but also its potentialities. Tattered words, unhinged from their representational ends in the domain of pure means are "blank," open to new "use" in testimony, born by a tear in future-present, while carrying the memory of woundedness.[78] In this aesthetic, as Beckett puts it, "language is most efficiently used where it is being most efficiently misused."[79]

As if in answer to the question, "how can [these nameless images] be represented," *The Unnamable* stages a series of "shabby, damaged" images. The first image on which I will focus alludes doubly to Max Nordau's book *Degeneration* and the camps; what is interesting is the ironic play between

Nordau's text, racial degeneration theory, and the "Final Solution." Beckett studied and took notes on *Degeneration* in the early thirties, the original notes for which still exist in Beckett's *Dream* Notebook.[80] In entry 613 of this notebook, Beckett scrawls the following, which he attributes to Nordau: "High degenerates, bordermen, mattoids, and graphomaniacs."[81] Any reader of Beckett will smile at the familiarity between these figures and the inhabitants of the Beckett country. Yet Beckett reported being unimpressed by *Degeneration*'s lack of tolerance, and felt his own growing thin.[82] Beckett may have made something like the degenerate body the protagonist of his *Three Novels*, but he performs an ethical reversal on Nordau's theory in *The Unnamable*. Here, the reference to Nordau is most notable in an almost deliberately allusive passage.[83] Mahood returns from abroad, where he has left his leg behind in Java: "its jungles red with rafflesia stinking of carrion." In Nordau's work, the bloody rafflesia—a parasitical plant—is a metaphor for the degenerate poet.[84] As Mahood hobbles home, he finds that his family has been "exterminated" by the sausage poisoning, another reference to Nordau, who makes reference to diseased society as poisoned sausage, while "extermination" is also suggestive of the systematic elimination in the camps.[85] The compound in which Mahood finds his family dead resembles a concentration camp. He finds himself in "a kind of vast yard or campus, surrounded by high walls, its surface an amalgam of dirt and ashes"[86] where the "screams of pain and wafts of decomposition" seem to him "quite in the natural order of things, such as I had come to know it."[87] With irony, once inside the camp walls Mahood exclaims, "I almost felt out of danger!"[88] The ruined allusion to the crematorium, where survivor testimonies report the smell of burning bodies was always palpable, is held in tension with the allusion to Nordau, whose eugenic theories, along with other racial socio-biological theory of the late nineteenth century, helped to prefigure Nazism.[89] In this juxtaposition, the narrative provides an ethical critique of the conditions that gave rise to the Holocaust.

Equally striking is the parody of the evidentiary language of factualism and testimony in the ensuing controversy over whether Mahood, once the narrator, and now the narrator's tormentor, has given him the "facts" about whether he reached his family before they died of the sausage-poisoning, and whether this "testimony" is reliable. The narrator wonders whether Mahood is telling the truth, as he questions "what really took place" or "what really occurred," as versions of the story are retold and discounted.[90] The testimony Mahood provides, like the pensum, resembles a disciplinary punishment, the way power forms subjects by "ramming a set of words down [one's] gullet."[91] Testimony here captures life instead of doing justice to it, its "face and profile, like [a] criminal."[92] Mahood provides "proof" for the Unnamable narrator, producing "ostensibly independent testimony in support of [his] historical existence," usurping the narrator's ability to speak.[93] The Unnamable must "testify to them, [his tormentors], until I die."[94] The debate over "testimony" as a factual record of what happened, the manipulation of this record by Mahood, and the ultimate failure of it to witness the lives of the drowned together raise the specter of the Holocaust. Primo Levi, in *The Drowned and the Saved,* remembers how SS militiamen cruelly enjoyed admonishing the prisoners that the record of what happened to them would be lost.[95] *The Unnamable*'s performance of the failure of representation, together with the ruined images of the camps, mediates on the ethical responsibility of testimony at the limit of fact. At the same time, it works to testify to the "secret" the Nazis sought to conceal.

Giorgio Agamben in *Remnants of Auschwitz: The Witness and the Archive* usefully theorizes these topics through a reading of survivor testimonies and continental philosophy. Dominick LaCapra rightly critiques aspects of Agamben's book for making use of the Holocaust as an instrument for something like the "negative sublime" in a manner that resembles the approach of Lyotard.[96] Beckett's work, however, provides an opening that allows us to read Agamben somewhat differently; that allows a reading of universal and contingent in the text that is less an appropriation of one by the other and more a shuttling of undecidibility between them. *Remnants of Auschwitz* in its theory of shame oscillates between the problems of history and structure in a manner that resembles Beckett's work. Building on survivor accounts such as Primo Levis' that narrate a splitting of the self between the lost and the saved, Agamben theorizes shame, or "disgrace," as a moment of "desubjectification," of being made conscious of "the disjunction between the living being and the speaking being that marks its empty place."[97] To feel shame is to "enter a vertiginous movement in which something sinks to the bottom, wholly desubjectified and silenced, and something subjectified speaks without truly having anything to say of its own ('I tell of things...that I did not actually experience')."[98] We will recall here that Beckett's *The Unnamable* begins with the prescient confession that "I shall have to speak of things of which I cannot speak," where reference sinks to the bottom in the vertiginous drowning of inscrutable life, and the imaginative faculties are forced to take over.[99] The ethical task in the aftermath of Auschwitz is to bear witness to the "drowned," those figures of bare life that embodied the horror of the inhuman in the human and who could not speak; for Agamben, the "witness" is the one that remains for the human whose humanity has been wholly destroyed.[100] The Holocaust victim is the only real witness capable of providing testimony; no one else can "speak for" the victim or claim to have lived through the victim's experience. However, this figure is unable to deliver testimony, inhabiting a non-place at the threshold between human and non-human, the non-space of inarticulation. Testimony in *The Unnamable* is produced where subjectivities are fractured, coextensive and at the same time non-coincident: "[the] subject, no longer there," and the one who can speak but is unable to speak, "a wordless thing in an empty place, [. . .] where nothing stirs, nothing speaks."[101] The diminished, porous body devolving gradually throughout *Three Novels* not only attests to the liminal space between the "I" and the "not I," the transhistorical traumatic meeting point where subjectification and desubjectification meet, but also the zone between the Shoah's drowned and the saved, it's victims and second-hand witnesses. Worm, Mahood, Basil and the Unnamable narrator inhabit separate subject positions and also exchange pseudonyms, alternate between master and slave, or come together as one: for Agamben, shame is found precisely in this region where subjectification and desubjectifiction coincide.[102]

The Unnamable stages an ensemble of radically disempowered beings that resemble the radically deracinated figure, the *Muselmann,* for whom Agamben writes *Remnants of Auschwitz.* As Agamben (and Primo Levi) detail, the *Muselmann* inhabited the lowest form at which life could exist in the Nazi concentration camps, one step away from *selektion.* At the threshold between life and death, the *Figuren,* or "doll," as the Nazis called the drowned, bore such abject suffering that Levi and Agamben have depicted this figure as beyond representation, as "speechless."[103] Early on, a seemingly wretched, disconsolate creature enters the narrator's realm of (a)perception:

> The other advances full upon me. He emerges as from heavy hanging, advances a few steps, looks at me, then backs away. He is stooping and seems to be dragging invisible burdens. What I see best is his hat. The crown is all worn through, like the sole of an old boot, giving vent to a straggle of grey hairs. He raises his eyes and I feel the long imploring gaze, as if I could do something for him.[104]

What is striking is the address of the figure in his long, imploring gaze, yearning to be perceived in his suffering. At the same time perception is denied, for the figure appears only through the grey metonyms of hat and hair like a black and white photo, lacking the fullness of representation or the vividness of color. Near the end of the novel an analogous gray haired other is rendered like an impressionist painting, a shadow or smear of himself, filtered through the narrator's curious multi-functioning eye that acts as both witness to and projector of images: "you set out to look for a face, to it you return having found nothing [. . .] nothing but a kind of ashen smear, perhaps it's a long grey hair [...] or all together, fingers, hair, and rags, mingled inextricably."[105] That the speaker sets out to perceive a face—a marker of human intimacy—but sees just an ashen smear and a mixture of fragmented parts goes to the heart of Beckett's anxiety surrounding "the remainder" in representation and "all that it blinds to," which confounds the factualism of testimony. Despite failure, these persecuted entities, like a stubborn stain, inhabit the page, a stain on the silence "like bodies in torment, the torment of no abode, no repose."[106] Testimony occurs in a zone of indistinction between the speechless and the speaking, the point at which neither "speaks." The drowned are bound to the saved, victim to observer, Worm to Mahood, Molloy to Malone to Murphy to their creator. The moment of articulation coincides with the moment of silence; testimony is borne where voice jams the machine of language. These broken images of the drowned address us as an apostrophe from which we cannot turn away.

The force of Beckett's text, as I have argued, lies in the very undecidability between the "transhistorical" and the "historical" dimensions of trauma. This account provides us with a way of reading alterity in Beckett beyond deconstructive celebrations of ineffable sublimity or displaced theology on the one hand, and humanist political projects that fail to nuance Beckett's impoverishment of the traditions of Western humanism (traditions in which Beckett was deeply immersed) on the other. While Beckett's syntax may be "tattered," not all semantic content is vacated, and we should look closely at the utterances and images in *The Unnamable* to read the twilight of an historical half-life faintly present, from which the text (de)generated. We can then register the ethical charge of *The Unnamable*'s aesthetic between absence and loss, inhuman and human, language and history.

Notes

I would like to thank Gregory Dobbins, Seán Kennedy, Dominick LaCapra, David Simpson, Bishnupriya Ghosh, Jonathan Hahn, Gerhard Richter, Patricia Moran, and the faculty and staff at the Beckett International Foundation at Reading for their generous assistance with this chapter. All errors are mine alone.

1. Slavoj Žižek, *Looking Awry* (Cambridge, MA, and London: MIT Press, 1992), 145.

2. See, for example, Slavoj Žižek's *On Belief* (London: Routledge, 2001). See Jean François Lyotard's *The Differend: Phrases in Dispute,* trans. Georges Van Den Abbeele (Minneapolis: University of Minnesota Press, 1988) and *Heidegger and "the jews,"* trans. Andreas Michel and Mark Roberts (Minneapolis: University of Minnesota Press, 1990).

3. Dominick LaCapra, *History in Transit: Experience, Identity, Critical Theory* (Ithaca and London: Cornell University Press, 2004), 121 n. Also see *Writing*

History, Writing Trauma (Baltimore and London: Cornell University Press, 2001) and *Representing the Holocaust: History, Theory, Trauma* (Ithaca: Cornell University Press, 1996).

4. Theodor Adorno, *Aesthetic Theory,* ed. and trans. Robert Hullot-Kentor (Minneapolis: University of Minnesota Press, 1997), 30.

5. Samuel Beckett, "Three Dialogues," in *Disjecta: Miscellaneous Writings and a Dramatic Fragment,* ed. Ruby Cohn (New York: Grove Press, 1984), 138–46.

6. LaCapra, *Writing History, Writing Trauma,* 188.

7. Ibid.

8. But see Gary Adelman, *Naming Beckett's Unnamable* (Lewisburg, PA: Bucknell University Press, 2004). Adelman suggests that *The Unnamable,* and the "spiritual excitement to which it gives rise" may correspond with Holocaust survivor testimonies. See also Jonathan Boulter's "Does Mourning Require a Subject? Samuel Beckett's *Texts for Nothing,*" *Modern Fiction Studies* 50.2 (2004): 332–50. Boulter engages concepts of mourning and trauma in Beckett's *Texts,* but argues that because Beckett's subjects are without history or memory, theories of mourning and trauma are ultimately unworkable. Also see David Houston Jones's recent, "From Contumacy to Shame: Reading Beckett's Testimonies with Agamben," in *Beckett at 100,* ed. Linda Ben-Zvi and Angela Moorjani (New York: Oxford University Press, 2008), 54–67 for a reading of the problematic site of articulation as testimony (Jones's article was not published at the time this article was composed so I am not able to give it more sustained reflection here).

9. LaCapra, *Writing History, Writing Trauma,* 105–6 and *History in Transit,* 137.

10. Samuel Beckett, *The Unnamable,* trans. Patrick Bowles (New York: Grove Press, 1958), 393. Trauma theorist Cathy Caruth argues that trauma is not fully assimilated as it occurs, and therefore it "defies and demands our witness." *Unclaimed Experience: Trauma, Narrative, and History* (Baltimore and London: Johns Hopkins University Press, 1996), 5.

11. Beckett, *The Unnamable,* 325.

12. Ibid.

13. UoR MS 3000, "*Whoroscope* Notebook," 84a. © Samuel Beckett Estate.

14. When Adorno suggests that Beckett's writing "hides its face," he refers to the presence in Beckett's work of bestialized (in)human beings that remain in life at the behest of instrumental forces. Theodor W. Adorno, "Trying to Understand *Endgame,*" *Can One Live After Auschwitz? A Philosophical Reader,* ed. Rolf Tiedemann and trans. Rodney Livingstone, et al. (Stanford: Stanford University Press, 2003). Beckett writes in the German Letter of 1937 that "more and more my own language appears to me like a veil that must be torn apart in order to get at the things (or the Nothingness) behind it." Beckett to Axel Kaun, 9 July 1937, trans. Martin Esslin in *Disjecta,* 171.

15. Adorno, "Trying to Understand *Endgame,*" 269.

16. Beckett, *The Unnamable,* 300.

17. Ibid., 375.

18. Adorno, *Aesthetic Theory,* 31 and *Negative Dialectics,* trans. E.B. Ashton (London: Routledge & Kegan Paul, 1973), 380.

19. Arthur Schopenhauer, *The World as Will and Idea,* vol. 1, book III., trans. Richard Burdon Haldane and John Kemp (London: Kegan Paul, Trench, 1907), 326.

20. For Schopenhauer's influence on Beckett, see James Knowlson, *Damned to Fame: The Life of Samuel Beckett* (New York: Grove Press, 1996), 248. Beckett to Thomas McGreevy, July 1930, Samuel Beckett Archives, Trinity College, Dublin, MS 10904.

21. It is in this context that he suggested "defunctus is a fine expression." See Arthur Schopenhauer's "Doctrine of Suffering" from *Parerga und Parapomena,* quoted in C.J. Ackerly and S.E. Gontarski, eds., *The Grove Companion to Samuel Beckett: A Reader's Guide to His Works, Life, and Thought* (New York: Grove Press, 2004), 431.

22. Ibid., 431.

23. Beckett, *The Unnamable,* 310.

24. Ibid., 311.

25. Ibid., 347.

26. Ibid., 369.

27. Ibid., 394.

28. LaCapra, *History in Transit,* 117.

29. Ibid., 116. I borrow the term "fictions of anteriority" from Gerhard Richter in "Acts of Memory and Mourning: Derrida and the Fictions of Anteriority," in *Mapping Memory,* eds. Susannah Radstone and Bill Schwarz (New York: Fordham University Press, forthcoming).

30. In an interview for the *New York Times* (May 1956), Beckett said, "In the last book—'L'Innomable'—there's complete disintegration....The very last thing I wrote—'Textes pour rien'—was an attempt to get out of the attitude of disintegration, but it failed." *Samuel Beckett: The Critical Heritage,* ed. Lawrence Graver and Raymond Federman (London: Routledge, 1979), 148. For poststructural and nihilist accounts of Beckett, see, for example, Anthony Ulhmann, *Beckett and Poststructuralism* (Cambridge: Cambridge University Press, 1999) and Shane Weller, *A Taste for the Negative: Beckett and Nihilism* (London: Legenda, 2005).

31. Beckett, *The Unnamable,* 291.

32. Beckett, "Three Dialogues," in *Disjecta,* 139.

33. Beckett, "German Letter of 1937," in *Disjecta,* 172.

34. Giorgio Agamben, *Profanations,* trans. Jeff Fort (New York: Zone Books, 2007), 88, 86. See also *Means Without End: Notes on Politics,* trans. Vincenzo Binetti and Cesare Casarino (Minneapolis: University of Minnesota Press), 2000.

35. Beckett, "German Letter," in *Disjecta,* 172. See Matthew Feldman's "'Agnostic Quietism' and Samuel Beckett's Early Development" in this volume for more detail on Beckett's relationship to the doctrines of quietism.

36. Beckett, *The Unnamable,* 330.

37. Ibid., 292.

38. Samuel Beckett, *Molloy,* trans. Patrick Bowles (New York: Grove Press, 1958), 149.

39. Schopenhauer, *The World as Will and Idea,* vol. 1, book IV, 358 n.

40. Beckett, *The Unnamable*, 348.
41. Jacques Lacan, *The Seminar of Jacques Lacan, Book II: The Ego in Freud's Theory and in Techniques of Psychoanalysis, 1954–1955,* ed. Jacques-Alain Miller, trans. Sylvana Tomaselli (Cambridge: Cambridge University Press, 1988), 154–55.
42. Beckett, *Molloy,* 81.
43. Beckett, *The Unnamable,* 293.
44. Ibid., 311.
45. Mahood/Worm's motto from Decartes is "De nobis ipsis silemus," ("we say nothing about ourselves"). Beckett, *The Unnamable,* 328–29.
46. Beckett, *Molloy,* 137.
47. Beckett, *The Unnamable,* 291.
48. Beckett uses the neologism "omnidolent" in "First Love." In *Samuel Beckett: The Complete Short Prose* (New York: Grove Press, 1995), 32.
49. Beckett, *The Unnamable,* 291.
50. Ibid., 296.
51. Ibid., 305.
52. Ibid., 297.
53. Ibid., 308.
54. Ibid., 304.
55. Ibid., 305.
56. Ibid.
57. Ibid., 306.
58. Ibid., 305.
59. Ibid., 306.
60. Ibid., 307.
61. Ibid., 411.
62. Jacques Derrida, "Jean-Marie Benoist (1942–90), The Taste of Tears," *The Work of Mourning,* ed. Pascale-Anne Brault and Michael Naas (Chicago: University of Chicago Press, 2001), 107.
63. James and Elizabeth Knowlson, eds., *Beckett Remembering/Remembering Beckett* (London: Bloomsbury, 2006).
64. Knowlson, *Damned to Fame,* 278–79.
65. Knowlson and Knowlson, *Beckett Remembering,* 79.
66. Ibid., 78 and 80.
67. Knowlson, *Damned to Fame,* 344–45.
68. Ibid., 291, 294, 301.
69. Knowlson and Knowlson, *Beckett Remembering,* 85.
70. LaCapra, *History in Transit,* 129.
71. Knowlson, *Damned to Fame,* 336; Knowlson and Knowlson, *Beckett Remembering,* 84.
72. Beckett, *The Unnamable,* 352.
73. Beckett, *The Unnamable,* 380.
74. Beckett, *The Unnamable,* 349. Pilling, *Beckett's Dream Notebook* (Beckett International Foundation, Reading, 1999), 35.

75. Ibid., 364, 383.

76. Ibid., 389.

77. Ibid., 407–8.

78. For more on tattered grammar, see Ann Banfield, "Beckett's Tattered Syntax," *Representations* 84 (Autumn 2003): 6–29.

79. Beckett, "German Letter," in *Disjecta*, 171–72.

80. William Inge, *Christian Mysticism* (London: Methuen, 1912). Pilling, *Beckett's Dream Notebook.*

81. Pilling, *Beckett's Dream Notebook,* entry 613.

82. Ackerley and Gontarski, *Grove Companion,* 408.

83. Beckett, *The Unnamable,* 317–18.

84. C.J. Ackerley, "Bloody Rafflesia, Coenaesthesis, and the Not-I," in *Beckett after Beckett,* ed. S.E. Gontarski and Anthony Uhlmann (Gainesville: University Press of Florida, 2006), 170.

85. Beckett, *The Unnamable,* 312. Also see, Ackerley, "Samuel Beckett and Max Nordau," 169.

86. Beckett, *The Unnamable,* 317.

87. Ibid., 322.

88. Ibid., 317.

89. Daniel Pick, *Faces of Degeneration: A European Disorder, c. 1848–1918* (Cambridge: Cambridge University Press, 1989), 27.

90. Beckett, *The Unnamable,* 320–23.

91. Ibid., 324.

92. Ibid., 362.

93. Ibid., 319.

94. Ibid., 324.

95. Primo Levi, *The Drowned and the Saved,* trans. Raymond Rosenthal (New York: Vintage, 1988), 11.

96. LaCapra, "Approaching Limit Events: Siting Agamben," in *History in Transit,* 144–95.

97. Giorgio Agamben, *Remnants of Auschwitz: The Witness and the Archive,* trans. Daniel Heller-Roazen (New York: Zone Books, 2002), 143.

98. Ibid., 120.

99. Beckett, *The Unnamable,* 291.

100. Agamben, *Remnants of Auschwitz,* 134.

101. Beckett, *The Unnamable,* 391, 386.

102. Agamben, *Remnants of Auschwitz,* 107.

103. Ibid., 120. For an important discussion of the "theatricality" of the play between Jew and Muslim in the *Muselmann* idiom (a play that Agamben misses), see David Simpson, *9/11: The Culture of Commemoration* (Chicago: University of Chicago Press, 2006), 160–67.

104. Beckett, *The Unnamable,* 298.

105. Ibid., 375.

106. Ibid., 391.

CHAPTER 6

Samuel Beckett, the Archive, and the Problem of History

Robert Reginio

"We speak so much of memory because there is so little of it left," writes Pierre Nora in his influential theoretical statement on the study of "sites of memory." Samuel Beckett's characters too speak so much about an existence that has been broken down to so little. The memory sheltered by modern institutions and archives, Nora continues, is "voluntary and deliberate, experienced as a duty, no longer spontaneous, psychological, individual, and subjective; but never social, collective, or all encompassing."[1] Memories seem to speak *through* Beckett's characters, who also experience their enunciation of the past as a duty or an obligation. This enunciation gestures towards some listener, but nevertheless fails to substantiate the social or the collective on Beckett's stage. The image of the archive, an image capable of encompassing the mythic and the modern, the logical and the fantastical, an image of the substantive and the perplexingly heterogeneous—this image and the conceptual discourse it forms is threaded through the plays of Samuel Beckett and their engagement with the problem of memory. The archive and the gestures of insatiable collecting and sheltering which inaugurate it are central to the modern conception of history, where the potentially disruptive presence of cultural fragments are transformed into the potential of a legible, purposeful "History" when encompassed by the shape of the archive. This shape implies archival work clears a space where, in the future, an objective perspective on historical events can be unproblematically situated.

Recovering the past in its totality, in its very physicality, seems to be behind the gestures associated with the archive, and suggests an attempt to limit the distance between the past and the present. Yet, the energy of this crossing-of-distances only reifies them: "Since no one knows what the past will be made up of next," Nora writes, "anxiety turns everything into a trace, a possible indication, a hint of history that contaminates the innocence of all things."[2] As an artist resolutely set against this utopian innocence, Beckett—through figures like the archivist Krapp—continually struggled to strip memories of their burden as traces, while recognizing that his drama inevitably instantiates, through its ironic subversions, the presence of the past. "Burning to be gone," it is nevertheless through a confrontation with his labyrinthine archive—the interleaved temporalities we hear emanating from his tapes—that Krapp seeks a way beyond the architecture of inscriptions that tacitly promises mastery over the past, but which disintegrates his (and the audience's) sense of his identity.

It is the gestural component of *Krapp's Last Tape*—the way it frames specific hesitations and interruptions through Krapp's manipulation of his recordings—which can direct a reading of the play engaged with a critique of the archive. And it is as a gesture that Jacques Derrida reads the archive. That is, for Derrida, the archive is not a neutral institution capable of generating objective perspectives on historical events. This lack of neutrality, or failure of objectivity, makes the very structure of an archive meaningful for Derrida. Since in his work dedicated to the archive (*Archive Fever*) Derrida draws his critique from a consideration of the historical development of psychoanalysis, the repression inherent in the archival gesture is not something to be overcome, but rather it is a symptom (of writing, in this case) to be placed at the center of our thinking about not only inscription but of *collecting* (contextualizing) those inscriptions *in time*. The implementation of an archive, collecting for the archive, and drawing narratives from that collection are acts which shelter and efface the history of the coming-into-being of those gestures. For Derrida, archival praxis and its attendant repression constitute the archive, and this repression shadows the narratives subsequently drawn from the archive. Derrida argues the archive literally and etymologically shelters "the memory of the name *arkhe*." *Arkhe*, in this context, refers to its Greek root *arkheion*, translated by Derrida as "initially a house, a domicile, an address, the residence of the superior magistrates, the *archons*, those who commanded... [who] were considered to possess the right to make or to represent the law."[3] Thus he defines "the law" (the meaning of certain events, we might say) not as something recovered *in* its texts, but as something contingent, something produced *through* the power of interpreting the archive's contents. The archive seems "to shelter itself and, sheltered, to conceal itself."[4]

In this chapter I will trace the image of the archive through two of Beckett's plays (*Krapp's Last Tape* and *Waiting for Godot*) and analyze how he dramatically disassembles the acts of concealment implicit in the archive. His drama places his characters beyond or in opposition to the "shelter" of the archive's structured embodiment of the past. Beckett's plays seem at key moments to exist only to negatively define the absences surrounding us. His drama is, however, filled with moments of resurgence; the dramatic tension is one between the evanescence of the actors' speech fading into silence and the utter necessity of the damaged bodies on the stage. This strategy is manifest in his reduction of dramatic means, itself a critique of representational theater. I understand this formal critique of Beckett's as, in part, a confrontation with some of the central crises of modern European history (namely exile and, more specifically, the Holocaust). By reading Beckett's drama in the light of his critique of the archive, I will examine how his work reveals the fact that these crises put into question the conceptual foundation of the modern archive and, in its extremity, human memory itself.

As a protean object of study and as an ever-shifting, oppositional concept used to critique traditional modes of narrating the past, memory currently occupies the center of the interconnected fields of trauma studies and Holocaust studies. By reading Beckett's drama through the lens of the Holocaust and the refinement of the concepts of "trauma" and "memory" in the work of writers who have been shaped by Holocaust and trauma studies, I intend to contribute to the critical reevaluation of Beckett's work as a response to (the problem of) history. It is Beckett's thoroughgoing critique of the archive in a play like *Krapp's Last Tape* that can serve as a conceptual bridge between his work for the theater and the pressing questions about history and memory at the center of post-Holocaust historiography and theory.[5]

Therefore, rather than seeing memory as a component of the fabricated bourgeois subjectivity unequivocally annihilated in Adorno's reading of *Endgame,* in this study memory—as an object and a process—remains equally resistant to categorization. Through these readings I will make explicit the exilic and post-Holocaust contexts of the plays and will highlight Beckett's resistance to universalizing rhetoric about history and memory. I will stress that Beckett's vexed figuration of memory on the stage suggests that this category, faculty, concept, or trace exerts a powerful pull on Beckett as an artist.

In Beckett's drama, the archive or the archival is rendered as an internalized, psychological, mnemonic realm that is frequently uncontrollable by the characters on the stage. The archival also appears in Beckett's minimalist drama as the external presence of ruins or fragments (even of theater itself) that are equally unmasterable. Beckett's characters repeat, replay, or reconfigure the internal mnemonic traces they carry. So unmasterable, these are not "memories" as commonly understood. Their memories collapse and are ultimately indeterminate once they are spoken in the theatrical space. In the face of this collapse, his characters recoil from these chaotic narratives and return to the harshly delimited space of their internal existence. As a trope equally suited to represent and understand both the psychology of memory and late modernity's conglomeration of external memorial traces or fragments, the archive becomes a site of refuge or an external threat depending on, in this case, the dramatic context.

Beckett Testifying against the Archive in Krapp's Last Tape

Neither individual nor collective, memory's ubiquity in modernity's philosophical obsessions and in its plethora of memorial traces and gestures signals an anxiety centered on the image of the archive. In *Krapp's Last Tape,* Beckett reimagines theater as the expression of this anxiety, where a fractured past creates an indeterminate future, a future indeterminacy that makes present experience duty-bound in every facet to collect and to re-collect. He does this by reducing the play's dramatic action from an actual dialogue between actors to the internalized dialogue between a present self and an externalized, archived past self. The archive on Beckett's stage consists of Krapp's collection of annual retrospective monologues recorded on reel-to-reel audio tapes. The dramatic action of the play consists of Krapp, aged sixty-nine and living a life of willful solitude, listening to his archived voice in preparation for his yearly retrospective recording. Each utterance heard on Beckett's stage in *Krapp's Last Tape* is meant to be heard at a later date; the future is implicated in, and thus makes contingent, present speech. In *Krapp's Last Tape* memories specifically voiced into a tape recorder in order to be listened to and understood later are "written" under the shadow of forgetting; they are inscribed in the face of an impending sense of oblivion. The danger for Krapp lies in separating the form of the archive from this potential meaning-making— within the very form of the archive, an intent is already at work.

Krapp's manipulation of his archive, his fast-forwarding or replaying of his memorial audio tapes, undoes the notion that new narratives can be brought to the archive in the hopes of building a present relationship with the past that excludes dominance of the present by the past or vice-versa. Beckett thoroughly undoes this notion of a neutral reciprocity between present and past by repetition, reduction, and dispersal. Krapp not only replays certain portions of his tapes, but his archive is constituted by a series of repeated gestures, the most basic of which being the annual, ceremonial process of recording. What is experienced as the archetypically singular performative moment for an audience—Krapp exposed within a beam of strong white light—is revealed to be the last of a series of repetitions as the play unfolds.[6] More incisively, Beckett *reduces* the dramatic engagement of self and archive (or "consciousness" and "history") to a conflict between present consciousness and personal memory, which in the play is materialized on audio tapes. The dramatic action of the play begins with Krapp's choice to reject love and a life within a community in favor of his *magnum opus*. His failure to construct this work satisfactorily is rendered on the stage through the occlusion of his present self, and by extension his very life, by the aggregate of traces embodied in his obsessively detailed archive, an aggregation which exceeds its purported end. Through this, an audience witnesses how his identity is *dispersed*. It is dispersed not merely across the reels of tape that house his reflections, but the fact that he has nothing left, that being itself—in the name of sheltering the long-lost creative "fire" at its core—has been materialized or archived on the tapes exposes what seems like either an inevitable failure of memory (the archival prosthesis subsuming Krapp's living memory) or the dominant ubiquity of memory. In any case, Beckett renders absurd the notion that knowledge can be gained through an engagement with the archive. But at the same time he gives us a figure in Krapp who can access being only through the archive.

At the end of *Krapp's Last Tape*, as Krapp records his final monologue into his tape recorder, he listens to his thirty-nine year old voice retell an incident of sexual consummation and an almost ego-less experience of another before he faces the audience with a silent, blank expression, the tape machine playing still, emitting not a sound. The tone of some of Krapp's last spoken words before this silent tableau ("Once was not enough for you") is a key to the entire play. Does Krapp rebuke himself here? Does he utter this as a bitterly salvaged scrap of wisdom? To answer this, a director or an actor must determine what precisely Krapp is doing in the final moments of listening. Is he trying, quixotically, hopelessly, to "be again"? Is he stunned or broken into the awareness of the chimerical power of the archival trace? Is this moment a rejection of archival limits or a surrender to them?

With Krapp, the audience struggles to naturalize the silence at the end of the play that had been populated by all those dead voices. The rest of the play sets the self-interrogation of an archivist as the context for the final scene of listening and silence. Early in the play, a thirty-nine year old Krapp speaking from the distance of the past captured on a tape, while trying to naturalize silence like many of Beckett's narrators, fills the present stage's aural emptiness with voices:

> *Tape*: Extraordinary silence this evening, I strain my ears and do not hear a sound. Old Miss McGlome always sings at this hour. But not

tonight. Songs of her girlhood, she says. Hard to think of her as a girl. Wonderful woman though. Connaught, I fancy. [*Pause.*] Shall I sing when I am her age, if I ever am? No. [*Pause.*] Did I sing as a boy? No. [*Pause.*] Did I ever sing? No.[7]

This moment encapsulates the basic structure of dramatic listening and the dispersal of the listening subject found in *Krapp's Last Tape*. A thirty-nine year old Krapp dramatically shapes silence in his monologue; Beckett presents us with a sixty-nine year old Krapp doing the same thing on the stage. It is the silent spaces within language—specified by Beckett in his stage directions—and the conclusion about the silences of Krapp's life offered on the tape that link the sixty-nine year old Krapp and his archive-bound younger self. When a memory is shaped by a present speaker through language, that speaker enacts a self-conscious recreation of self, the inaugural phrase "I remember…" containing the essential theatricality of the self-conscious enunciation.[8] At this moment, Krapp on the stage is situated between the materialization of the past and the present tense of performance. The silence which putatively clears the ground for introspection is flooded with memories—this clearing of the ground itself ("Extraordinary silence this evening.") is a memory inscribed on the tape. The older Krapp's own silence on the stage is shadowed by the taped memory. Introspection is complicated by the way the archived voice "stages" or theatrically shapes a moment of silence in which the self is supposed to be unequivocally present to itself. The self, in both instances, is displaced by memory.

On the tape, the voice of a younger Krapp invokes "Connaught" in the silent void. "Connaught" is a place name, and as such it is the sound of community and presence. More so than other types of words, including personal names, place names are a sort of linguistic archive. They carry the genealogy of a place or—through translation, anglicizing, or renaming—the trauma of conquest or dislocation. Like so many moments in Beckett's drama, a setting of no determinate place or time (Krapp resides in his "den" in the "future") is contaminated by the pathogen of history and memory. In this play, the linguistic archive of "Connaught" disrupts the stylized *mise-en-scène* with its geographical and historical specificity.

A place name, like a fragment set in an archive, is evidence of a life lived in a larger historical continuity, and yet without the context provided by living memory and all of its contingencies, it is, like any signifier, something empty, reliant on the differences of lived history if it is to mean anything at all. A voice archived gains its meaning specifically through the structure of institutions contextualizing it or the location of those who "read" them. Miss McGlome—herself the harbinger of the "gloaming" of mortal twilight—has failed in her domestic ritual during the evening in which the younger Krapp makes his recording. Into this absence Krapp sketches her history, using the opportunity to wonder about his own life measured in the formal schema of childhood, growth, and old age. This is a gesture not unlike Krapp's entire archival project, an archival project that eventually seems an impediment to "be again." An archive can enable repetition—the "again" in Krapp's imperative "be again"—but it cannot vouchsafe being itself. This dual movement between silence and speech, between the fragment as evidence and the fragment as empty signifier, between listening and interruption, and between being and memory's repetitions constitutes the tensions which drive Beckett's drama.

More specifically, the tension between the textuality the archive implies (a fixity that nevertheless raises the problem of repetition) and the presence of Krapp on the stage (a presence the title tells us is at its end) can perhaps be seen as the inaugural conflict from which all of the rest follows. As Krapp finally speaks into the recorder near the end of the play, he voices a desire to recoil from the performance, to turn to a fixed textual world. The sixty-nine year old Krapp describes his limited forays over the past year outside his "den":

> Crawled out once or twice, before the summer was cold. Sat shivering in the park, drowned in dreams and burning to be gone. Not a soul. [*Pause.*] Last fancies. [*Vehemently.*] Keep 'em under! [*Pause.*] Scalded the eyes out of me reading *Effie* again, a page a day, with tears again. Effie… [*Pause.*] Could have been happy with her, up there on the Baltic, and the pines, and the dunes. [*Pause.*] Could I? [*Pause.*] And she? [*Pause.*] Pah![9]

The world of texts spills over into the world of Krapp's memory and vice-versa. Along with the "girl in a shabby green coat,"[10] the "dark young beauty" of a nurse,[11] his mother, the girl in the punt, and the "old ghost of a whore" Fanny, Effie Briest—a fictional character—is another woman that focuses his concentration, and she is another woman he eventually rejects. He reads Theodor Fontane's novel *Effie Briest* almost like he records, (supposedly) limiting himself to a page a day, as Krapp limits himself (supposedly) to one annual archival entry. His interaction with Fontane's novel, and perhaps its nature as a text with a fixed (and fictionalized) beginning and end, mirrors his interaction with his archive. The only difference is that the novel is bracketed off as a fiction while his archive offers up the possibility that the "grain" of his existence can still be separated from its husk.

"Burning to be gone" from an existence he has purposefully de-socialized and rendered solitary, Krapp turns to Fontane's text specifically because of its iterability. Opposed to these "last fancies" however, Krapp dismisses the delimited text. And yet Krapp on the stage, and the play's audience, comes to this material as a trace, as a fragment drawn from his archive—any gesture of rejection made in the name of a reduced existence, in this play, is enveloped by the archive. Memory's continual materialization in the shadow of the archive reveals itself not as trace nor as substrate nor as a "signified," but as this continuous dissatisfaction, as the embodiment of the past in an iterable form in the name of the singular, the non-iterable. Krapp's evocation of Effie is a microcosm of the play's conclusion, when Krapp rejects the order of his archive—ending his present recording—in order to grasp and perhaps even relive the memory of the enigmatic, romantic moment spent with the girl in the punt that is inscribed on the tape. And it could be that Krapp the exile is grasping after the historical specificity of that moment—the promise of a lived continuity enfolded in "Connaught," for example—as he sits in a space drained away to its elemental black and white, light and shadow, speech and silence.

Beckett prepares his audience for the play's equivocal, imagistic ending by making Krapp's interjection of gaps, silences, hesitations, and deformations of the archival text into the dramatic action predominate, establishing a rhythm of movement and stasis, of silence and the aural archive's "speech." Krapp has been trying to recall the fleeting moment of enigmatic communion with the girl by the lakeside. He has a vague notion of where his description of this moment is located on the tapes and must, through a series of associations and work with a ledger, discover the passage indirectly. And this is of a piece, since Krapp does not want to actually locate the episode in a chronological system or pattern. He does not want to understand it as origin or as endpoint, and in fact the hesitations and various textual deformations performed by Krapp resist, without denying, the final stage image as the dramatic action's culmination. Krapp does not want the understanding the structure of the archive lends the memory, a structure which houses, nevertheless, the voice of the past. As Krapp listens to a spool of tape, searching for the description of his encounter with the girl in the punt, the taped voice continues on after describing the death of Krapp's mother:

> *Tape*: Spiritually a year of profound gloom and indigence until that memorable night in March, at the end of the jetty, in the howling wind, never to be forgotten, when suddenly I saw the whole thing. The vision at last. This I fancy is what I have chiefly to record this evening, against the day when my work will be done and perhaps no place left in my memory, warm or cold, for the miracle that . . . [*hesitates*] . . . for the fire that set it alight. What I suddenly saw then was this, that the belief

I had been going on all my life, namely—[*Krapp switches off impatiently, winds tape forward, switches on again*].[12]

Like Krapp, listening to this opaque passage, the audience is waiting for the archived voice to yield a moment of meaning. It is a remarkable moment in which hesitation is dramatized as both a temporal pause and a psychologically meaningful refusal to speak. The revision the thirty-nine year old Krapp makes after the hesitation for which the stage direction calls—correcting himself and saying "the fire" instead of "the miracle"—is a significant rejection of abstraction in favor of a more concrete word. Both Krapp listening and the voice of Krapp on the tapes are trying to unearth something concrete—a tense hesitation mirrored in the audience's expectation. Our adherence to narratives that project an archivable "meaning" is evidenced by our frustration at the old Krapp switching off at the threshold of the revelation of his "miracle."

If the ultimate form of the archive is one that sets the stage for a later construction of a comprehensive narrative, then in *Krapp's Last Tape,* it is through hesitation, deformation, effacement and an ultimate rejection of the archive that Krapp's self as we confront it on the stage is constituted. The rejection of the archive in *Krapp's Last Tape* is an instance of dispossession as much as it is a gesture of resistance. "Memory works through characters" on the postmodern stage, writes Jeanette Malkin, "but never originates in them."[13] The physical body takes on a troubling presence on a stage defined by its non-referentiality. "Beckett not only presents the body as an object to be manipulated through the mechanisms of discipline and control," insists Anna McMullan, "but also focuses on the body as…a space or dynamic which eludes or resists linguistic or specular control."[14] Karen Laughlin goes as far to suggest that Krapp continually returns to the memory of the girl in the punt because "that encounter exceeds representation and the documentary impulse associated with it."[15]

The paradoxical coupling of the documentary impulse and Krapp's desire to resist it can be found in the very act of speaking into the archival machinery. We hear the confession of the archivist in the very first tape Krapp plays. Listening to entries he recorded years earlier—his "post mortems"—seems not to sharpen his memory, but their failure to record anything of value raises his ire and seems to spur him on in the present. The thirty-nine year old Krapp on the tape says: "These old P.M.s are gruesome, but I often find them—[KRAPP *switches off, broods, switches on.*]—a help before embarking on a new…[*hesitates*]…retrospect."[16] A hesitation in the past is framed by a brooding hesitation in the present. The first pause, as Krapp switches off the tape recorder, suggests that Krapp is either still thinking about "Bianca on Kedar Street" (what the thirty-nine year old Krapp at that time consid-ered "hopeless business") or that Krapp himself is considering his archival project. The reader of the play also hesitates between the specific historical references in the play ("Kedar Street," *Effie Briest,* Connaught) and Beckett's dissolution of the archival processes which bring these references to light. The hesitation of the thirty-nine year old Krapp recorded on the tape under-scores the paradox of the entire play, a play whose present action is utterly consumed by the past. Krapp is at once ending and "embarking on a new retrospect"; he speaks finally, but he is always speaking with the intent to listen once again in the future. If his memory is built upon or wedded to the archive of the tape, then the territory of the past is both utterly circum-scribed or not at all.

The trope of memory or the past as a terrain in which to wander or a territory to map seems obvious, but when we see that Krapp's present or taped speech takes place within a dramatic structure consonant with that of testimony—testimony being a speech act self-consciously performed for the archive (literally "for the record")—we can note the greater complexity with which Beckett handles this traditional *topos* of "inward journeying." We assume there is a body of events, an as-yet-unnarrated story out there which the archive will discover. The impetus which drives us to record is one and the same with that which assumes there is a stable foundation from which an individual speaks. To speak *into* a recorder is thus to admit contingency in the same (literal) breath as one speaks *as* an indissoluble self. Beckett bril-liantly uses the tape recorder to dramatize this moment where the absurdity of being emerges, its fictionality and its ineradicable presence.

Tabori's Waiting for Godot: *The Post-Holocaust Dimensions of Beckett's Drama*

The coupling of the documentary impulse—the archival impulse—and the exposed body on the stage in *Krapp's Last Tape* echoes the structure of post-Holocaust testimony, where both the impulse to resist the comprehensiveness of the archival inscription co-exists with (and raises questions about) "mem-ory" and "history." Although this play is not itself a piece of post-Holocaust testimony, it nevertheless stages the tension between the purported ends of the archive—a juridical conclusion to an historical narrative—and the nec-essarily excessive presence of the witness and the unending gesture of enun-ciating his testimony. George Tabori's 1984 production of *Waiting for Godot,* by explicitly dramatizing the archival aspects in and of Beckett's text, specif-ically reflects the problem of testifying about catastrophes that in themselves reconfigure testimony, rendering the act of witnessing an endless task.

The son of a Hungarian Jew murdered in Auschwitz, George Tabori was one of the millions of exiles produced by Europe's cataclysmic twentieth century. As playwright and director, Tabori went on to merge the two dominant strands of post-war European theater: Brecht's historically situated meta-theatrical gestures and Beckett's more abstract, autonomous theater. In Tabori's writing and directing, this fruitful intersection is defined by the problem of repetition. Where Brecht replays actual historical moments in the present (the Renaissance in *Galileo* or the Thirty Year's War in *Mother Courage*) in order to have his spectators take an active, ironic stance towards the narration of history, Beckett's characters engage in a purgatorial repetition of personal memories, the significance of which remains veiled to the audience and to themselves. And his most successful interpretation of Beckett's themes through Brechtian theatrical practice (or his interpretation of Brecht through the themes of Beckett) was his production of *Godot* in Munich in 1984.

Tabori's production of *Waiting for Godot* in Munich preceded, by approximately two years, the *Historikerstreit* in Germany, when the question of collective memory pressed upon historians and the German nation as a whole. The question of repetition, as a theoretical category and as a cultural phenomenon, asks about the role of history in culture. When is a focus on the Holocaust a debilitating repetition? When does a discourse that speaks of the future and reconciliation nevertheless repeat the past? More generally, and in contrast to narratives of reconciliation, is the early modernist focus on narratives of temporal cleavage and radical, almost utopian new forms commensurate with post-Holocaust reality? More specifically, should Europeans resign themselves to repeating the themes of guilt as they "work through" the past, or should a critique of the past be mainly set against the threat of repetition? Repetition becomes both the shape of accounting for the Holocaust and the shape of a subconscious phenomenon that threatens a true understanding of the past.

Repetition is an undeniable thematic and dramatic centerpiece of Beckett's text. However Tabori's production gives this theme new life and a poignant particularity. In a cultural and political climate where questions of repetition played a central part, Tabori—an artist specifically engaged with the post-Holocaust problems of memory—produced Beckett's play as an exploration of how drama, as a particular form of art, manages, appropriates, embodies, and ultimately suffers repetition. He revisits the theme of repetition in a thoroughly Beckettian manner; that is, Tabori takes a modality particular to drama and lends it ambiguous metaphorical weight (as in the "lighting director" of *Play* or *Catastrophe*'s "rehearsal"). He staged *Godot* as a "performed rehearsal," the actors dressed in contemporary clothes struggling, text in hand, to perform the play and yet failing to use this performance, a skilled execution of their art, as a springboard towards making meaning in the world beyond the theater. By bringing the text of *Godot* to the stage and by dramatizing the rehearsal process, Tabori mobilizes an archive, and the discourse of the archive, on the stage. The key to Tabori's innovative production of Beckett's *Godot* is his making explicit the archival substratum of the performative moment. Each moment of effective performance, in this production, is shadowed by the archive, by the presence of the play-text on the stage. If the relationship between the repetition of the past and the rhetoric of transcendence is an emblem of the problem of memory in the modern era, then the modality of drama—of a text's relationship to a lived performance—is a discursive complex well-suited for the expression and exploration of this problem. In modern, self-conscious theatrical experimentation repetition becomes the shape of accounting for the problem of memory as well as a disruptive force to be contained by dramatic form.

Tabori's "performed rehearsal" of *Godot* was a meta-theatrical gesture wherein the struggle to embody a text, with the aims of transcending its semantic limits, parallels the struggle of Beckett's characters to fashion an existence beyond the bounds of their endless waiting.[17] The actors wear modern day clothes, drink coffee, smoke cigarettes, and carry their well-worn copies of the play-text to the stage. A director, producers, and assistants sit outside the "rehearsal" space, observing. Jonathan Kalb describes how the actors playing Estragon and Vladimir, at the start of each act "enter languidly in what could be slightly aged street clothes and read indifferently from their scripts [until] they gradually work up to performance tempo, leaving their scripts for longer and longer intervals."[18] Physically entering the present time of performance, the actual text of the play exhibits the destabilizing effects of the archive in Beckett's theater. Some of the stage directions from the text were read out by the actors during the production and they referred many times to the text itself—sometimes with a sense of security and also with a sense of claustrophobia completely in line with the play's exploration of the debilitating effects of repetition.[19]

Just as the characters' enigmatic obligation to wait structures and orders their time on the stage while at other times serving to constrict their existence, so the text, literally present on Tabori's stage, is both refuge and obstacle. For example, the first time the actors voice the famous refrain "Let's go. / We can't. / Why not? / We're waiting for Godot. / Ah!" the actor playing Vladimir holds up his copy of the play-text as he says "We're waiting for Godot," signaling that for both the actors and for the characters those actors are representing, present actions are circumscribed, are dictated, are repetitions. When, alternately, the characters look back to the past and debate the events that have transpired, the actors physically gestured towards the text, suggesting their search for certainty. Physical movement on the stage was accomplished only when the actor had seemed to memorize his lines, feeling secure at that point to leave his text on the table. The process of "becoming" their characters, the process of moving from the scripted word to the spontaneity of lived speech, only to be reminded of how this "spontaneity" is shadowed and determined by a text—this indeterminate process parallels the situation of the characters in the play. The script is a paradoxical source of and impediment to "authentic" speech. Tabori's production, then, resists a reductive allegorization of Beckett's play as it "historicizes" it. In a cultural context where disturbing, catastrophic events were being inserted into larger historical narratives in order to bring their debilitating effects under control, Tabori's production questions the archive of a script as either origin or ideal culmination. By extension, he asks if any work of art should be endowed with such "archival" qualities.[20]

By figuring the dramatic text as an archive—a source for the actors, the origin of the dramatic action, and the horizon of the possible on the stage—Tabori touches on an essential struggle present in Beckett's work for the theater: a struggle between memory figured, on the one hand, as an ineluctable presence, a presence marked by its disappearance and return, a presence characters continually try to lend a voice, and, on the other hand, memory as always unsubstantiated, as in *Godot,* or memory as something to be resisted, such as in *Not I.* As an institution and as a theoretical concept, the archive is both a source of and a tomb for memory. At moments in Tabori's production, the characters seem determined to speak spontaneously, to give voice to their damaged memories unencumbered by the tomb-like text-archive. However, the *rejection* of the archive on Beckett's stage is a never completed task.

As in any survey of Beckett's *oeuvre,* rejections signify beginnings, not endings—rejecting the armature of traditional illusionistic referentiality, Beckett begins his work for theater. Similarly, the rejection of the archival inaugurates a series of repetitions, rather than embodying the theatrical gesture as such, as it is for Artaud. The basic structure of *Godot* is premised on repetitions, pairings, twinnings, and theatrical speech which is circumscribed as well as meta-theatrical. In order to comprehend their paradoxical existence, the characters proffer histories and genealogies only to observe them dissolve into the same bare story. Dialectical movement (between, say, Pozzo and Lucky as residual carriers of a master/slave relationship and Didi and Gogo as epigones, both slaves and masters in an unstructured social space) collapses into relativistic sameness. "The circularity of theatrical perception," writes Anthony Kubiak, "finds a kind of phantom completion in [these doublings endemic to Beckett's drama]: post-apocalypse becomes prehistory."[21] Of course, what's elided or out of reach here is "history" itself—something potentially accessed through the archive. The archive cannot be defined unequivocally as either end-point or site of inauguration and thus cannot be rejected *in toto*. In other words, the past is ever-threatening the present in Beckett, specifically because the past cannot be articulated. The archive is always both "post-" and "pre-"; it is always equally the resting place of history's fragments as well as a site from which new histories can be made.

The *obligation* the characters have to this archive, the way they are bound to the play-text on the stage in Tabori's production (the obligation of *performance*) is never framed in determinate, ethical terms, and yet the presence of this obligation exists undeniably for Beckett's characters; it is as essential to the drama as is the reduced, ruined landscape of the plays. Where does the difference fall exactly between Lucky's obligation to speak (made manifest at the end of Pozzo's whip) and Vladimir and Estragon's obligation to their colloquy and ultimately to Godot? Tabori's *Godot* production implies theater itself is intimately marked by this obligation to speak, to replay, and to repeat.

Reading Beckett's drama from the theoretical perspective of Derrida's deconstructive analysis of the archive and the dramatization of speaking in the shadow of archive found in both *Krapp's Last Tape* and Tabori's dramaturgy illustrate several points of contact between Beckett's art and the problem of memory as it is understood after the Holocaust. Beckett's drama gives voice to the pressures of post-traumatic memory because the very foundation of the plays is the failure of language to account for a catastrophe that nevertheless shapes the dramatic landscape.[22] But, since this catastrophe cannot be accounted for in language, its coordinates remain painfully indeterminate. In this way, one can say that the struggle of Beckett's characters to master language (i.e., to delimit the coordinates of the catastrophe that defines their existence) is the struggle of the traumatized victim to master traumatic repetition.[23]

Tabori's production of *Godot* casts the characters as European exiles, intellectuals caught in a dislocated space that could be perceived as either post-apocalyptic or pre-historic. The specific historical context of Tabori's production, however, urges us to see this state of "permanent exile" as emblematic of the post-catastrophic, post-Holocaust status of historical knowledge. Tabori imagines theatrical space as a site where various discursive and memorial energies intersect and overlap. Theater in this case is both historically situated (i.e., self-consciously theatrical) and formally autonomous (i.e., non-mimetic). In a study of theatrical representations of the Holocaust, Edward Isser writes that

> Tabori is one of the few theater artists who has managed to synthesize the political and historical ideas of Brecht with the imagistic and ritualistic work of Growtowski and Beckett...[he] creates dramatic scenarios that meet Adorno's criteria for authentic post-Holocaust representation.[24]

I take this to mean that Tabori meets Adorno's criteria for "authentic" post-Holocaust representation because, as in his *Godot* production, his characters debate their own authenticity. I would concur with Isser in that his production helps us to ask what constitutes an "authentic" production of Beckett, or of any play. Therefore, in terms of the archive, we can ask if authenticity is achieved when a production occludes its archival origins, life quite literally being breathed into the puppet-like roles potential in a text. Yet we can also ask if a production can be considered authentic when it integrates an uncompromising self-awareness into performance, as in Brecht's *Verfremdungseffekt*. Tabori's production links Beckett and the Holocaust in an evident, literal way—by producing *Godot* in a Germany coming to grips with the Holocaust, we can conjecture if Tabori was making Beckett testify for the catastrophe. If this is indeed true, Tabori makes the play testify for the catastrophe, as Kalb points out, in a way that remains true to the self-reflexive non-representational aspect of all of Beckett's drama. More so, Tabori's production draws upon the play's thematic and formal engagement with the archive, an engagement which serves as a bridge between the play and its specific historical context.

The absurdity of an existence that seems both post-apocalyptic or pre-historic is embodied, in *Krapp's Last Tape,* in the archive that contains silence. A text cannot contain silence, but a tape-recorder can. The machinery of inscription and its gesture towards historical specificity are, at this moment in Beckett's play, set against themselves. It can record in silence and can accumulate aporias, as it does in Beckett's play, but the gesture of the archive—its gesture towards capaciousness and subsequent comprehensibility—is set against silence. Post-Holocaust historiography and testimony are defined by the tension between these gestures as well; the fidelity to silence and the incommunicable aspects of the catastrophe co-exist with the archival impulse. As we have seen, this particular tension is enwoven throughout Beckett's dramatic evocation and concomitant deconstruction of the archive. Fragments in an archive are supposed to speak with an almost seductive authority. Beckett's plays, however, can contain silence and make it meaningful, showing how silences are also constitutive of the self, how forgetting is constitutive of remembering. Only through interaction and interpretation do silences become meaningful: they have to be mobilized, they have to be staged.

Tabori's production of *Godot* figures the performative moment as that when memory is spoken out of an archive by an individual whose very existence as an autonomous being depends upon a rejection of that archive. Memory voiced out of a figurative, internal, psychological archive or voiced from a material, textual archive inevitably points to the absences which encircle it. And it is only through these vital absences that testimony finds its meaning in a post-Holocaust Europe. Beckett is preeminently concerned in his formal experimentation with letting these absences reveal themselves for an audience. The irreplaceability of the drama in revealing these absences suggests its testimonial power.[25]

Notes

1. Pierre Nora, "Between Memory and History: *Les Lieux de Mémoire,*" *Representations* 26 (Spring 1989): 7, 13.
2. Nora, "Between Memory and History," 17.
3. Jacques Derrida, *Archive Fever: A Freudian Impression* (Chicago: University of Chicago Press, 1998), 2.
4. Derrida, *Archive Fever,* 3.
5. For an indication of how Beckett's literary work might intersect with the current debates in Holocaust and trauma studies, see Dominick LaCapra, *History in Transit: Experience, Identity, Critical Theory* (Ithaca: Cornell University Press, 2004), 187; "Holocaust Testimonies: Attending to the Victim's Voice," in *Catastrophe and Memory: The Holocaust and the Twentieth Century,* ed. Moishe Postone and Eric Santner (Chicago: University of Chicago Press, 2003), 222; and *Writing History, Writing Trauma* (Baltimore: Johns Hopkins University Press, 2001), 23.
6. Exteriorization—whether in writing or in the construction of an archive such as Krapp's—opens up a space for repetition. There is no archive without repetition, or, as Derrida explains, "there is no archive without consignation in an external place which assures the possibility of memorization, of repetition, of reproduction, or of reimpression." *Archive Fever,* 11.
7. Samuel Beckett, *Krapp's Last Tape,* in *The Complete Dramatic Works* (London: Faber and Faber, 1990), 218.
8. The "theatricality" of memory results from it being spoken to another. This specific definition of "theatricality" is at the heart of Maurice Halbwachs' theory of memory. The past "is reconstructed on the basis of the present" and the present is shaped by societal, familial and institutional pressures (what he terms "collective memory" and "social frameworks for memory"), so much so that, for Halbwachs "it is to the degree that our individual thought places itself in these frameworks and participates in this memory that it is capable of the act of recollection." Maurice Halbwachs, *On Collective Memory* (Chicago: University of Chicago Press, 1992), 38–40. Halbwachs' theory has not been applied to Beckett's work,

but his notion of memory as constituted by a series of listeners and speakers is not entirely distinct from Beckett's work for the theater. Not just in *Waiting for Godot* and *Endgame,* but in his later drama such as *Not I* and *Ohio Impromptu,* memory invariably takes the shape its enunciation and collaborative reformulation bestows it.

9. Beckett, *Krapp's Last Tape,* 222.
10. Ibid., 218.
11. Ibid., 219.
12. Ibid., 220.
13. Jeanette Malkin, *Memory-Theater and Postmodern Drama* (Ann Arbor: University of Michigan Press, 1999), 7.
14. Anna McMullan, *Theatre on Trial* (New York: Routledge, 1993), 11.
15. Karen L. Laughlin, " 'Dreaming of […] Love': Beckett's Theatre and the Making of the (Post)Modern Subject," in Angela Moorjani et al., eds., "Endlessness in the Year 2000/Fin sans Fin en L'an 2000," *Samuel Beckett Today/Aujourd'hui* 11 (Amsterdam: Rodopi, 2001), 207.
16. Beckett, *Krapp's Last Tape,* 218.
17. Antje Diedrich describes the production in this way: "The play was staged in the round with the enclosed performance space defined by a round area of sand, diminished to the outlines of a circle in Act II. Two joined tables surrounded by eight chairs stood in this arena. A black working light hung above the table with a black wire dangling from it, which turned into a green, leafy twig after the interval: a remnant of Beckett's tree." "Performance as Rehearsal: George Tabori's Staging of Beckett's *Waiting for Godot* and *Endgame,*" in Mathijs Engelberts et al., eds., "Historicising Beckett/Issues of Performance," *Samuel Beckett Today/Aujourd'hui* 15 (Amsterdam: Rodopi, 2005), 148.
18. Jonathan Kalb, *Beckett in Performance* (Cambridge: Cambridge University Press, 1989), 92.
19. "At first, the points when they drop their scripts seem like sections that the actors happen to have memorized…Later those sections seem like improvisations, the actors attempting to depart from the text, or at least to lose themselves sufficiently in the action that they may call it 'theirs.' What holds them is the routine the script dictates, which they cannot escape no matter how hard they work towards spontaneity; for they must return tomorrow, speak again the very same words, and try to make that activity seem worth doing." Kalb, *Beckett in Performance,* 91.
20. The dramaturgical conceit of the "staged rehearsal" balances both the abstract language of Beckett's dramatic practice and Tabori's exploration of how memory unfolds in a historically specific present. Anat Feinberg, writing on a similar production of Beckett's work by Tabori, notes that the "energy of the production derived from the unresolved tension between antithetical qualities: the abstract or formalistic character of the Beckett oeuvre and Tabori's attempt to render it concrete," resulting in "an open-ended experience: not only was waiting a never-ending occupation for Vladimir and Estragon, so was the acting—or

rehearsing—of the play." Anat Feinberg, *Embodied Memory: The Theatre of George Tabori* (Iowa City: University of Iowa Press, 1999), 132, 142.

21. Anthony Kubiak, "Post Apocalypse Without Figures: The Trauma of Theater in Samuel Beckett," in *The World of Samuel Beckett*, ed. Joseph H. Smith (Baltimore: Johns Hopkins University Press, 1990), 109.

22. In the specific context of film, trauma, and the Holocaust, Joshua Hirsch argues that trauma is not a thing, like a letter, that can be delivered: "[r]ather, trauma, even before being transmitted, is already utterly bound up with the realm of re-presentation." Trauma is "a crisis of representation," the events themselves "neither properly stored for recall, nor able to be forgotten." Joshua Hirsch, *Afterimage: Film, Trauma, and the Holocaust* (Philadelphia: Temple University Press, 2003), 15. Krapp's memory of the girl in the punt is "improperly stored," subject to the limits of archived language, but his desire for this lost love cannot be forgotten. Expanded to encompass temporal and geographical distances, Krapp's struggle with the archive expresses the trauma of exile.

23. Ruby Cohn identifies the theatrical convention of soliloquy or soliloquizing as a threat for many of Beckett's characters. Soliloquy threatens the character with a situation not unlike that besieging the testifying victim of trauma: "As long as [the main characters of *Waiting for Godot*] can volley words, they stave off soliloquy, which may engulf a man in solitude." *Just Play: Beckett's Theater* (Princeton: Princeton University Press, 1980), 39.

24. Edward Isser, *Stages of Annihilation: Theatrical Representations of the Holocaust* (Madison, NJ: Fairleigh Dickinson University Press, 1997), 110.

25. If the drama is to fully engage with history, memory, and the archive, it therefore seems essential that theatrical practitioners find ways to evade the strictures put upon Beckett's dramatic texts by the Beckett Estate and to build upon Tabori's powerfully suggestive production that is now, alas, left to the archive. For an engagement with the "archive" as a way around these strictures see S.E. Gontarski, "Reinventing Beckett," *Modern Drama* 49.4 (Winter 2006): 419–27.

CHAPTER 7

Archives of the End: Embodied History in Samuel Beckett's Plays

Jonathan Boulter

The subject has died—and perhaps many times—along the way.

Samuel Beckett, *Proust*

My life, my life, now I speak of it as of something over, now as of a joke which still goes on, and it is neither, for at the same time it is over and it goes on, and is there any tense for that?

Samuel Beckett, *Molloy*

Does anything remain?

Samuel Beckett, *Happy Days*

Introduction

Samuel Beckett's characters are haunted by the ghost of memory. Living in a protracted state of regret and loss over what the speaker of *A Piece of Monologue* calls the "one matter. The dead and gone,"[1] the Beckettian subject cannot ever seem to move past the claims, the remains, of history. In his 1917 paper "Mourning and Melancholia," Freud distinguishes between mourning, which he argues is the "normal" way loss is comprehended and worked through, and melancholia, the pathological response to loss which situates the subject in a continual position of narcissistic identification with the lost object. The melancholy subject cannot accept loss and works pathologically "to establish an identification of the ego with the abandoned object."[2] History—loss, trauma—continually works its way into the present moment because the subject cannot or will not move past the traumatic moment. The melancholy subject maintains this pathological state—which, as Eng and Kazanian argue in *Loss,* may not be so pathological after all— because the traumatic moment is a kind of definitional threshold articulating the subject's sense of (traumatized) interiority: the subject, in a sense, *becomes* her trauma.[3] In *Proust,* Beckett argues that the past threatens continually to "deform" the subject and alerts us, as he will do in his drama, to his awareness of how lethal, yet unavoidable, a continual and thus melancholic, relation to history can be: history is not something that has passed but is "irremediably part of us, within us, heavy and dangerous."[4]

Part of the startling effect of Beckett's drama is the way in which this thematic of history and melancholia is instantiated. Unlike some of the prose which delineates what may be purely mental or ideal spaces—think here of the curious spaces of *The Unnamable* or *Texts for Nothing*—the body is an insistent presence in Beckett's drama: despite or because of his attempts to reduce, fragment, or displace it, like the repressed, the body returns to become a repository, an archive, for the remains of history.[5] The character may exist (as in *Happy Days,* or *Not I*) in a radically diminished physical state at what appears to be the end of human culture, but she—specifically, her *body*—maintains the archival traces of that lost culture and becomes what Pierre Nora calls a "prosthesis memory."[6] The images Beckett gives us in his drama—a disembodied mouth, people in urns, a body immobilized in a wheelchair or the earth—are images that concretize the idea of embodied or archived memory, of subjectivity and thus history inextricably linked, cathected to, shards of the material.

The Beckettian archive, however, does not, as Derrida argues in *Archive Fever,* mark the space of beginnings or of futurity, necessarily. The Beckettian archive is always one marking the end, tracing what he understands to be the remainders of cultural signifiers. And yet, as the repeating structures of these plays indicate, the archive of the end never does fully end. Winnie's parasol will reappear, Hamm will re-play his various narratives, Mouth will be compelled to speak again: the Beckettian subject who becomes the embodiment of what I term the "melancholy archive," is thus always a historical subject or, more precisely, *subject to,* a repeating and thus aporetic history. This repetition signifies the ceaseless inevitability of history's claims and thus becomes a perfect formal expression of Freudian melancholia. In readings of *Happy Days, Not I,* and *Footfalls,* my essay will work towards an understanding of the idea that Beckettian archive—which becomes a trope for Beckettian interiority—is fully dependent on the compromised—and repeating—body which in its turn demarcates what can only, and paradoxically, be understood as the continual, *melancholic,* end of history. If the archive is understood as a means by which to organize the past and to orient an understanding of a spectral futurity, if, that is to say, the archive serves to organize a progressive temporality and thus orient an understanding of interiority itself, it is my purpose here to suggest that Beckett's representation of the *continually ending* archive works as a critique both of temporality and interiority and thus ultimately of the archive's foundational premises.

The Archive and the Specter

The archive, as Derrida argues in *Archive Fever,* is oriented to, just as it is defined by, a peculiar structure of temporality. We perhaps instinctively conceive of the archive as a response to the past, as working towards a preservation of what has been, but Derrida wishes to emphasize the degree to which the archive more accurately works authoritatively to mark out the space of beginnings and futurity. He draws our attention to the etymology of archive noting that the term itself means a place of commandment and commencement: the archive thus operates to assert a kind of juridical control:

> *Arkhe,* we recall, names at once the *commencement* and the *commandment.* The name apparently coordinates two principles in one: the principle according to nature or history, *there* where things *commence*—physical, historical, or ontological principle—but also the principle according to the law, *there* where men and gods *command, there* where authority, social order are exercised, *in this place* from which order is given—nomological principle.[7]

Note Derrida's insistence on the what we could call spatial claims of the archive: it is a place—*there, there,* he insists—of geographical, phenomenal, reality; it is a place from which the order of things—let us call it for now the order of history—is governed. I wish to emphasize Derrida's emphasis on the spatial reality of the archive: it is a space, a site, a phenomenal presence just as it becomes a temporal and spectral entity. Beckett's archive too is phenomenal and spectral, real and historical: an embodiment of (spectral) history.

The archive, for Derrida, marks a space of anxiety, precisely an anxiety about the possibility of loss: the archive exists only as an anticipation (and we note the futurity of this concept) of the loss of history; as such it works proleptically to preserve what will inevitably be lost. The temporal valence of the archive thus is precisely futural: "the archive takes place at the place of originary and structural breakdown of the said memory."[8] Derrida speaks of the technical archive (and we may think here of Krapp's recording machine): "the archive...is not only the place for stocking and for conserving an archivable content *of the past*...the technical structure of the *archiving* archive also determines the structure of the *archivable* content even in its very coming into existence and in its relationship to the future."[9] It is precisely here, in the anxious relation to what *will be,* that the spectrality of the archive comes into play: I am going to offer two quotations here:

> It is a question of the future, the question of the future itself, the question of a response, of a promise and of a responsibility for tomorrow. The archive: if we want to know what that will have meant, we will only know in times to come. Perhaps. Not tomorrow but in times to come, later on or perhaps never. A spectral messianicity is at work in the concept of the archive and ties it, like religion, like history, like science itself, to a very singular experience of the promise.
>
> [T]he structure of the archive is *spectral.* It is spectral *a priori:* neither present nor absent "in the flesh," neither visible nor invisible, a trace always referring to another whose eyes can never be met.[10]

In Derrida's complex reading of Yerushalmi's reading of Freud—in this way setting *Archive Fever* up as a kind of archive of Yerushalmi and Freud—the specter of Freud is instantiated (like the arkon) as an authority, but a spectral authority, a phantom to which Yerushalmi (and Derrida) orient their own temporal positions of coming *after:* after the Father, indebted to the Father, working perhaps to assert their own archiving gesture of authority over the figure who cannot be effaced. This "impossible archaeology"[11] of return can only figure history within a logic of the spectral inasmuch as the origin is both past but uncannily, and insistently, always returning as the trace (of the trace). We note here how Derrida shifts from emphasizing—at least etymologically—the phenomenal reality of the archive (it is a *place, there*) to a reading which insists on its inevitable spectrality. The archive is neither present nor absent in time: it perhaps will never be present; it is no longer, nor ever was, a material entity, neither present nor absent *in the flesh.* But it is precisely the embodied quality—the fleshiness—of the archive, of history, that Beckett seems to assert—or from which he cannot escape—in plays like *Happy Days, Not I,* and *Footfalls.* Beckett's archives compel us into conceiving the archive as at once flesh and specter, body and ghost: the hauntology of the Beckettian archive is one which must be able to negotiate a (perhaps aporetic) reading which *sees* the body—this is after all, *theatre*—as a material presence upon which the specter of traumatic history is traced.[12]

It is here that I wish to suggest a reading of the Beckettian body, which is still yet an archive, through the trope of the crypt. We can, to begin, return briefly to Derrida and his reading of melancholia in *The Ear of the Other*. Here, drawing on the work of Abraham and Torok, Derrida offers an image of melancholia which is uncannily fleshy; it is an image of history as a kind of viral, material presence, working its way into the body of the melancholy subject; it is an image, I wish to suggest, particularly suited to a reading of the Beckettian archive:

> Not having been taken back inside the self, digested, assimilated as in all "normal" mourning, the dead object remains like a living dead abscessed in a specific spot in the ego. It has its place, just like a crypt in a cemetery or temple, surrounded by walls and all the rest. The dead object is incorporated in this crypt—the term "incorporated" signaling precisely that one has failed to digest or assimilate it totally, so that it remains there, forming a pocket in the mourning body. The incorporated dead, which one has not really managed to take upon oneself, continues to lodge there like something other and to ventrilocate through the "living."[13]

I am interested in exploring how the dead object—which here assumes the force of history and of personal loss—asserts itself upon the subject who herself is always already spectral: that is to say, we need to understand how the dead object is taken up—encrypted—within the melancholic subject who is articulated into a "world" of pure and unending repetition. If, as I will suggest, the specter is primarily a trope for repetition, for *absolute* return— and we need to keep in mind precisely the degree to which a character like Winnie, for instance, is placed within a ceaseless, almost machined, world of repetition—how can we begin to conceive of the specter who is *herself* haunted, by history, by loss, by what Winnie calls the "old style"? How, precisely, is Beckett asking us to conceive of his spectral, yet embodied, subjects? Just what are we to make of the insistent presence of the body here?

Perhaps the best way to begin answering these questions is to call to mind the material reality of the Beckettian performance, the fact that the melancholia of Winnie, of Mouth, of May, takes place, is "acted out" on stage.[14] If I am suggesting that we begin thinking of the body on stage as a kind of crypt containing the traces of a history which will not be worked through, I am also suggesting, perhaps insisting, that the stage itself operates as well as a crypt. The stage space—whether the *"Very pompier trompe l'oeil"* of *Happy Days*,[15] the uncanny, seeming non-space of *Not I* (but it is only a

seeming non-space), the confined and confining space of the hall floor in
Footfalls—works as a material container of the histories being played out,
works, more precisely, as a space within which the traces of histories are
mapped out onto the body of the grieving subject. There is thus a doubling
of the crypt in Beckett's theater: the body as crypt, the stage as crypt. What
I am interested in emphasizing here is that the theatricality of the crypt—
the play as play—has as much to do with communicating the idea of history as the body which bears the traces of history's passing: there is, thus,
a reciprocal exchange of spectralities between body and stage in Beckett's
hauntology. If the stage in some ways thus is always a trope for memory,
the body—precisely as it is staged, as it asserts itself in a space to be seen,
to be witnessed by the audience—becomes memory, embodies memory as
it speaks for memory, *in memory's place.* But this speaking for memory, of
memory, works always in Beckett at the level of absolute disavowal: Winnie
insists on continuing on, in maintaining a stiff upper lip in the face of the
absolute waste of history plainly in front of her; Mouth similarly is unable,
unwilling, to speak of memory as her own. The crypt, as Derrida reminds
us, is a space where things are hidden ("it hides as it holds"):[16] if the body
is not hidden in Beckett's stage, if the body is rather an insistent presence,
what is disavowed, what moves towards being absolutely disavowed, is history as conjoined to the subject-as-body: history itself is encrypted in these
plays precisely to the degree that it is hidden in plain sight.

Embodying the Crypt: Happy Days

The now iconic image of Winnie as she appears in *Happy Days*—"imbedded"
up to above her waist in act one; to her neck in act two—is of course an
image of burial, of encrypting. Winnie is a crypt, encrypted: she is a repository for history's end, an archive for the flotsam and jetsam of history,
encrypted to the point of near annihilation. Her strikingly uncanny position immediately calls for interpretation and, indeed, Beckett anticipates
this. Winnie offers a narrative (perhaps it is a memory, but she seems to
be inventing the story as she goes) of being observed by a Mr. Shower (or
Cooker) who in his turn asks the questions the audience is asking: "What's
she doing? he says—What's the idea? he says—stuck up to her diddies in
the bleeding ground—coarse fellow—What does it mean? he says—What's
it meant to mean?"[17] Our readings of Winnie are made difficult, are compromised perhaps, not only by the strangeness of her image, but by the way
the image—and the totality of the play—seems at once to be allegorical and
real, and hence uncanny in almost classical fashion. That is to say, one of the
difficulties of reading Winnie is that she seems at once to be so obviously a

symbol of entrapment—she is trapped, say, within the claims of a decayed marriage; she is trapped within the claims of a falsely soothing fetishized capitalism ("There is of course the bag...The bag...The depths in particular, who knows what treasures...What comforts");[18] trapped and disabled by history for which she has become the mouthpiece—and a real woman, *really* trapped for some reason and by some unknown means in the earth after (perhaps) some cataclysm.[19]

Because Winnie is fully present as a woman, as a body, as a subject: "*About fifty, well preserved, blond for preference, plump, arms and shoulders bare, low bodice, big bosom, pearl necklet.*"[20] Beckett's "well preserved" may be a casual use of a sexist metaphor or it may simply be a true statement: she is indeed well preserved by the crypt she inhabits, preserved enough to feel the limits and pains of her embodiment: "The earth is very tight today, can it be I have put on flesh, I trust not."[21] And it is here, within this body, within this archive of the end, that various histories are in turn preserved, well or otherwise. We begin with Winnie's recitation of a prayer which ends in the surely ironic "World without end Amen":[22] the world has indeed seemingly come to some kind of end, but Winnie is placed within a context of absolute repetition (I will return to this point) which absolutely vitiates the concept of the *heavenly* infinite. Here she is "no better, no worse" because here there is "no change."[23] Here she has only the remains of a culture—toothbrushes, mirrors, handkerchiefs, parasols—to comfort her; here perhaps she herself preserves the remains of a defunct culture by insisting on using these products in the "old style." Winnie, moreover and perhaps most crucially, becomes the repository of misremembered lines from literature; if she herself is well preserved, the remnants of her literary culture surely have not fared as well: Winnie misremembers—remembers ill, we can say—lines from *Hamlet, Paradise Lost, Romeo and Juliet,* Gray's "Ode"; yet, crucially, lines are quoted (from Herrick; Yeats; Keats; Milton; Shakespeare) which are entirely appropriate to their context,[24] indicating that Winnie is functioning not simply as a repository for non-contextualized fragments of a culture long passed, but that she is able, perhaps in a quasi-Arnoldian fashion, to draw on these touchstones in order to find what Beckett (in *Watt*) calls "semantic succor."[25] This is to say that Winnie, at one level—and it is, granted, yet only one level—functions as a traditional archive, preserving what has been lost, preserving the history of a culture.

But our question must become: what kind of archive can function within the specific temporality Winnie inhabits (or is inhabited by)? At one level Winnie marks the end of a culture; precisely, she functions as the marker of the end of a culture, the end of the old style of things, the order of things (she inhabits a time and space where all life has seemingly vanished: the appearance of an emmet—like the flea in *Endgame*—is cause for incredulity: "Looks like life of some kind!").[26] To speak of the end of things, however, is to beg the question of temporality itself: it is, precisely, to assume time functions. Even, perhaps especially, Derrida's conception of the spectrality of the archive, of its absolute futurity—"It is a question of the future, the question of the future itself"[27]—is one articulated by a (conservative, certainly Western because messianic) sense of temporality as a linearity; the archive works to answer a question posed *now* to be answered *then*. But Beckett's archive, Beckett's cryptic archive, does not function within the logic of a linear temporality (and this will be true of *Not I*, as it surely is of *Waiting for Godot* and *Endgame*). Winnie will at once acknowledge what appears to be a conventional, historical, that is, linear progression of time—"I speak of when I was not yet caught—in this way—and had my legs and had the use of my legs, and could seek out a shady place"[28]—and to cancel the very possibility of conceiving of historicity as allowing the event to emerge separate from the machinery of repetition: "Yes, something seems to have occurred, something has seemed to occur, and nothing has occurred, nothing at all, you are quite right."[29] In this world, this post-temporal, posthumous (perhaps post-human: I will revisit this idea in my conclusion) world, the very idea of event—as singular occurrence—can only ever be an absurdity:

> The sunshade will be there tomorrow, beside me on this mound, to help me through the day. (*Pause. She takes up mirror.*) I take up this little glass, I shiver it on a stone—(*does so*)—I throw it away—(*does so far behind her*)—it will be in the bag tomorrow, without a scratch, to help me through the day. (*Pause.*) No, one can do nothing. (*Pause.*) That is what I find so wonderful, the way things... (*voice breaks, head down*)...things...so wonderful.[30]

What Beckett presents us here, through Winnie, is a world articulated by absolute loss—the world has ended, this is a wasteland—but in which *nothing can be lost*.[31] In such a world, a world without the possibility of the event, the possibility of the event of loss, there cannot be any working archive as such. "May one still speak of time?" Winnie asks.[32] And of course, in a world of absolute machined repetition, the answer can only ever be no. As time is vitiated, as loss becomes impossibility, Winnie becomes an archive of absolute repetition, marking a never-ending stasis; she stands in relation to the absolute loss of a world from which she can never turn, can never escape. Precisely, she cannot mark that loss as loss because it continues *in her*. If the archive is articulated, grounded on the idea of loss, on the idea that it marks the space of loss—perhaps marks the space of an anticipation of loss—Winnie, as body, as subject, can only ever mark the space where loss is never comprehended as such because no distance is possible between her own interiority and the loss that defines that interiority. Here "[t]here always remains something…Of everything";[33] here, remains function always as spectral reminders not only of a world that once was, but shall ever be only always spectral. These things—and Winnie herself surely *embodies* one of these things as much as her mirrors and parasols—"have their life, that is what I always say, *things* have a life,"[34] but it is a life marked by the nullification of the possibility of ending and, as such, is not a life at all: "that is what I *always* say."

Perhaps we need to tighten the focus here: if there is no possibility of the event as such—if, that is, the event requires a temporality to be understood as event—can we not say (following Winnie's own words) that we have, *at least*, the event of nothing? "Nothing has occurred," she says, and perhaps we should take her at her word. The event of nothing, which comes to resemble the event of absolute repetition, surely does become the event of the spectral, the spectral event. "[T]he only way one can speak of nothing is to speak of it as though it were something," to borrow again from *Watt*;[35] "Nothing is more real than nothing," we read in *Malone Dies*.[36] Winnie maps out her life, marks out her relation to absolute loss, as a subject, as an embodiment of loss: as such she does become a phenomenal marker of absence as such. If nothing has occurred it occurs from within the space of an embodied response to nothing, to the passing of what *was once* time. And it is here that we arrive at the crux of the Beckettian archive, its paradoxes, its aporias: loss can only be traced—shall we say, archived?—by the remains of that loss, by the traces of that loss. Loss marks itself, remarks on itself, traces itself and as such, preserves its own absence, its own event of nothingness. Perhaps the easiest way to express this paradox is that loss must always, for Beckett, be marked by a body, if only a trace of a body; loss must always find its cathexes in a body which traces that loss as an absolute hold even on itself. Loss is encrypted, in other words, in the space of its own absolute disappearance.

Bearing the Voice: Not I and Footfalls

Questions of loss, its disappearance—or disavowal—are central to *Not I* and begin to point to the play's uncanny expropriation and transposition of tropes of the crypt, of the archive-as-crypt. *Not I*, visually, presents us with an extreme image of the subject as crypt *within* a crypt. *Not I* presents an image of the subject—Mouth, fragmented, reduced to a speaking role—within the encompassing blackness of a space—or is it a representation of temporality, a time of nothing?—which operates as a visual emblem of the crypt, that space operating as an enclosure upon the dead, the hidden. And Mouth is in a sense buried here in this (stage) space: the curtain which rises as the beginning of the play does so to reveal her as always already there, repeating—seemingly endlessly—the story, the history (but perhaps we need to call it a fantasy) of her trauma. Beckett's use of the mouth—beyond the obvious pun on not I/ not eye—speaks to another sense of the crypt, another enclosure from which history speaks: the mouth is both the crypt and the vehicle by which an encrypted history is made known even as it is, crucially, disavowed: "what?.. who?..no!..she!"[37] It is my purpose here to think through the implications of Mouth's story, to ask questions about the subject's relation to her history, to ask, precisely, what happens to the archive—and the archive of history— when that history is disavowed. My suggestion here—and I am following the lead of Abraham and Torok—is that disavowal works always within the economy of fantasy: Mouth's story, her primary narrative of the assault (if that is indeed what it is) on that April morning, serves only as a mask for some other entirely different melancholy, which here is to say, an entirely different *history*. The assault has served this one purpose: it disinters her history, compels her to voice a past which has remained encrypted and hidden. This process, too painful to bear (the voice of history in Beckett is always too painful to bear), becomes the means by which a history can be voiced and at the same time disavowed, spoken and retracted. This process, finally, is one which must be embodied insofar as it is in fact given voice: history, in other words, must find its lost object—the body.

And indeed *Not I* is all about the process by which something, some history, is revealed, brought to light even in defiance of the subject's desires. Behind the curtain, before it rises, we can hear Mouth "unintelligible"[38] mumbling what we can safely assume to be the story she is always telling. Mouth, it seems, is condemned—"obliged" might be the better Beckettian word—to speak her story endlessly as if, perhaps, being punished for some crime. And certainly Mouth has an acute sense of guilt, a sense that her assault on that April morning may have been some divine retribution: "her first thought was...oh long after...sudden flash...brought up as she had been to believe...with the other waifs...in a merciful...[*Brief laugh*]...God...[*Good laugh*]...first thought was...oh long after...sudden flash...she was being punished.... for her sins...a number of which then..."[39] But what guilt is possible here? What crime is possible here? In some sense that crime in Beckett is being born, is, perhaps more accurately, being witness to being improperly born into this world:

...out...into this world...this world...tiny little thing...before its time...in a godfor—...what?..girl?..yes...tiny little girl...into this...out

into this...before her time...godforsaken hole called...called...no
matter...parents unknown...unheard of...he having vanished...thin
air...no sooner buttoned up his breeches...she similarly...eight months
later...almost to the tick...so no love...spared that[40]

Mouth's world—can we yet call it an interiority?—from its inception,
her conception, is one marked by absence, by the phantom: the vanished
father,[41] the mother, dead, or simply gone. And we notice how these absences
are traced by tropes of speech and voice: the parents are "unheard of"; she
is spared the love that is "normally vented on the...speechless infant."[42]
Mouth's early history is marked by silence, by the unheard, that her forced
recitation, the play itself, in effect cancels. Perhaps the guilt that infuses
Mouth's sense of herself, the guilt that thus infuses her language, is the
guilt of speech itself, the guilt of being compelled into a world of language.
Where before she has been "practically speechless...all her days,"[43] now her
voice—but she cannot admit it is her own—works as a conduit of a lost,
because unspoken, history "dragging up her past".[44] her past, unearthed,
is that which before has remained hidden, encrypted, because far from language. Language now reveals the past, disinters the past, and confers an
interiority at once cryptic (but can we avoid the spatial, *cryptic* metaphor
in the notion of interiority any longer?) and disavowed: "when suddenly
she realized...words were—...what?..who?..no!..she!...realized...words
were coming...imagine!..words were coming...a voice she did not
recognize...at first...so long since it had sounded...then finally had to
admit...could be none other...than her own."[45]

In relating the narrative of her fall into speech, Mouth eight times uses
the word "imagine": words were coming, she says, "imagine!";[46] she "gradually realized...she was not suffering...imagine!";[47] she has "no idea...what
she was saying...imagine!";[48] she feels "her lips moving...imagine!";[49] she
senses feeling coming back into her body (her "machine," as she refers to
it): "feeling was coming back...imagine!";[50] she says where before she never
spoke, "now can't stop...imagine!"[51] At one level the word for Mouth is
simply a rhetorical reflex meaning "What an odd occurrence! Fancy that!"
But Beckett's composition is careful here and we notice how he conjoins
the word to her repeated narrative of that April day. In one sense Mouth is
asking the audience—or some interlocutor, her own self—to imagine this
event; at a more crucial level, I suggest, Mouth is signaling to herself the
need to (re)imagine, to re-invent that day. In Beckett's world there is often
the suggestion that imagination and invention are not dissimilar acts of
mind.[52] I am interested in the rhetorical valence of this word, its suggestion
that Mouth is asking herself to imagine, forcing herself perhaps, back into

that moment of the entry into speech: in some sense there is a desire—or at least an inability to avoid, which may be the same thing as desire—to return nostalgically to that moment. But I also wonder if the word "imagine" does not here function to signal not only the compelled return to a spot of time, but a fantasy of a return, a phantasmal return: this is to say, in Mouth's melancholia—and recall that I define melancholia following Freud as the inability, perhaps the unwillingness, to let the past lie, to let the past be the past—surely some element of fantasy is at work, fantasy, as Abraham and Torok would suggest, that masks the real trauma, the real substance, that is, of Mouth's interiority.[53]

Mouth has become, by her own admission, only a part of herself: "whole body like gone . . . just the mouth . . . lips . . . cheeks . . . jaws . . . never . . . what? . . . tongue? . . yes . . . lips . . . cheeks . . . tongue never still a second . . . mouth on fire."[54] Reduced to a speaking part, a part of speech, but never of course *apart* from her speech—she has become in a sense, language itself, language's compulsion—Mouth claims for herself a narrative explaining the etiology of her forced entry into the Symbolic: isolated before from the social because of her lack of speech, and where her occasional lapses into speech were only causes of shame ("then die of shame"),[55] Mouth has existed only for herself. Her fall into language, as she indicates, is a traumatic entry into discourse which forces her to confront her own history, the Symbolic register, shall we say, of her own life. Mouth's entry into speech is, in fact, the opening of the archive of herself, an opening—I have used the metaphor of *disinterment* before—entirely painful: "begging the mouth to stop . . pause a moment . . and no response . . . as if it hadn't heard . . or couldn't . . couldn't pause a second . . . and the brain . . . raving away on its own . . trying to make sense of it . . . or make it stop . . or in the past . . . dragging up the past . . ."; her thoughts lead inevitably back to her birth: "or that time she cried . . . the one time she could remember . . . since she was a baby . . . must have cried as a baby . . perhaps not . . . not essential to life . . . just the birth cry to get her going . . breathing."[56]

I would suggest that Mouth's real agony occurs when memory is compelled upon her forcing her back to the moment of her entry into world. The trauma of her fall into language precipitates the trauma of returned memory: her language puts her, as she states, "in the past." And the phrase that haunts this play, more so than the screams of denial, is the quietly—but three times repeated—"no love . . spared that."[57] This phrase, so reminiscent of the emotional register struck by Krapp in his own memory work, indicates the real absence in Mouth, the real "godforsaken hole."[58] Thus Mouth, before her fall into language, before her entry into spoken memory—and Beckett's play seems to suggest that memory is only memory as it is spoken, as it is given voice (Mouth never indicates she has no memory before the fall, but that only she never has had occasion to speak of them)—has lived something of an idyll, never having to confront what surely is a past of absolute misery. Abraham and Torok speak of "endocryptic identification" in their essay "The Lost Object—me"; the subject has created a split in the psyche, created in fact "a crypt in the ego,"[59] where an originary loss or traumatism is hidden from itself; to mask this originary trauma the subject must create a fantasy to which blame for present misery, melancholia, can be attached: "The melancholic's complaints translate a fantasy—the imaginary sufferings of the endocryptic object—a fantasy that only serves to mask the real suffering, the one unavowed, caused by a wound the subject does not know how to heal."[60]

In my reading of *Not I*, Mouth's assiduous denials that she is the one suffering the fall into language works to expose her birth and subsequent desertion by parents as the originary trauma that cannot be faced, cannot be voiced save through the intricate mechanism of archived, and phantasmal, melancholia. It is Mouth's injunction—to her interlocutor, her lost object, herself—to *imagine* that speaks to the possibility that the trauma of that April day is nothing but a fantasy trauma which works masochistically to return Mouth, perversely, to the scene of her originary entry into the world. Here we have a mechanism similar to what Freud, in *Beyond the Pleasure Principle,* noticed in his analysis of traumatic neurosis, specifically the way his patients' dreams kept returning the traumatized subject to the original scene of agony. Freud was compelled to theorize that either trauma was such a powerful psychic mechanism as to disrupt the function of dreams or, ominously, "we may be driven to reflect on the mysterious masochistic trends of the ego."[61] The process by which Mouth's memories, are disinterred, unearthed, by the fantasy of attack, speak clearly to the idea that, in Beckett, the past can never be disavowed: despite the fracturing of the subject, despite that is, her reduction to a psychically split metonymic aspect of herself, memory persists despite the subject, perhaps *to spite* the subject.

Not I is one of Beckett's most complex analyses of the relation between memory and interiority, between the archive and the self. At one level memory is only ever the memory of a trauma which in its turn disinters the real past of the subject, the real past which, too painful to face, to name, to speak, nevertheless is revealed even as it is denied. Thus, for Beckett, interiority—the crypt in which the subject is hidden even from itself—is coterminous with the archive; the archive—the storehouse of memory—only ever stands in disavowed relation to the subject and as such, as it is disavowed, the archive, like the subject, can only be discontinuous, can never rise to the status, as theorized by Derrida, as place of commandment and commencement. What we can propose thus is the idea that for Beckett the archive of memory—call it history as well—is always exteriorized even as it constructs a discontinuous interiority in the subject: the archive must always remain (as remains) "out there," beyond the subject, preceding and exceeding the subject. And yet because, again *for* Beckett, subjectivity and interiority always must be materially realized, must always be embodied, the archive must be voiced, must, at least—and here we turn to *Footfalls*—be spoken.

Of the plays considered here, *Footfalls* most obviously thematizes the idea of the archive as specter. The woman's voice ("*from dark upstage*")[62] is fully disembodied (it reminds us of the woman's voice in *Eh Joe*) leading us to suspect, perhaps, that she is dead; May herself, while a presence onstage, nevertheless is decidedly spectral, with her grey hair, grey clothing, her voice "*low and slow throughout*";[63] when the lights come up at the end of the play, there is no trace of May indicating, perhaps, that she was only a specter, a ghost now vanished.[64] Moreover, this is a play about the past, a history which cannot easily be linked up to the subject on stage: that is to say, the story May tells in her "sequel,"[65] the story of Mrs. Winter and Amy both is and is not the story of May and her spectral Mother. May's story of Amy is thus a spectral version of the story of her own life: with the transposition of one letter, "Amy" for "May," Beckett draws attention to how easily displaced the narrative of one's life can be, how very easily one's life, the narrative of one's life, can be lost.

And surely the idea of losing one's life in another, for another, perhaps *because of* another, is the theme of this dreadfully melancholy play. May has, it seems, linked her life so closely to that of her Mother as to have lost any claim to a life of her own.[66] As the play opens with what appears to be a dialogue between May and her Mother (it may very well be a dialogue between May and the memory of her Mother, May and the ghost-voice of her Mother), it becomes clear that May has devoted her life, her forty years, to caring for her Mother: "Would you like me to change your position again?...Change your drawsheets?...Pass you the bedpan?...The warming–pan?...Dress your sores?"[67] The story the voice tells of May beginning her life of pacing makes clear that May has renounced a life in the world for a life inside what will become a crypt of sorts:

> Where is she, it may be asked...Why, in the old home, the same where she...the same where she began...Where it began...It all began...But this, this, when did it begin? When other girls of her age were out at...lacrosse she was already here. At this...the floor here, now bare, once was.[68]

It becomes clear, as we register the mysterious absence of the body of her Mother, that May, precisely, has become a crypt for memory, for history: when the woman's voice announces "I walk here now"[69] before telling the story of the origins of May's pacing, May becomes a vehicle for her Mother's voice, a body through which, upon which, the Mother will speak herself into a continual relation to the living (if indeed May is alive). May's interiority, now fully cryptic, now fully the archive for the other, is effaced in a perfect representation of melancholic incorporation. Derrida argues that successful mourning affects a kind of killing of the dead other: the lost other (what Abraham and Torok call the "lost object") is taken into the body: "I take the dead upon myself, I digest it, assimilate it, idealize it, and interiorize it in the Hegelian sense of the term...I kill it and remember it." But in melancholia "I cannot interiorize the dead other so I keep it in me, as a persecutor perhaps, a living dead."[70] What is clearly at stake in Derrida's reading of melancholia—and what is clearly at stake in *Footfalls*—is identity itself. Mourning allows the subject to kill and remember the past, that is, to assert some kind of agency over it; melancholia is a form of possession where one's subjectivity is defined by one's relation to what is taken within and *remains* (versus what is taken within and digested). May is deformed by her mother, just as she is defined by her: "*I walk here now.*"[71]

And thus when May herself speaks, and speaks of herself in the third person, she distances herself from herself, ghosts herself: "A little later, when as though she had never been, it never been, she began to walk...At nightfall."[72] May displaces herself into the story of Amy and Mrs. Winter, doubling in uncanny fashion—but the uncanny is always about the encounter with the ghosts, the simulacrum—the story of her own life. And we note the pathos here in terms of May's assertion of herself as story-teller: if her agency, her interiority, has been displaced by that of the improperly incorporated Mother, here she attempts to reclaim a kind of authority through narrative itself. But what kind of agency does a *specter* have? What kind of agency can May, who tells the story of herself (as Amy), assert when what is being "revolved," in endless circular return, is a narrative of displaced subjectivity, a narrative essentially figuring May as subject to the dominance of her Mother? One of the uncanny effects in this last section of the play is how May's assertion of narrative, in which she speaks for her Mother (in the second section the Voice of the Mother is heard exterior to May), collapses into itself precisely as a failed assertion of authority. May does indeed appropriate, speak for, her Mother here, but given the degree to which (in the play's second section) May is dominated by the personality of her Mother, we have no sense of power being asserted by May: she speaks for her Mother, speaks as her Mother but the effect is a doubled authorizing of the mother's discourse, rather than May's own. And when, finally, May speaks as her Mother speaking to Amy, and when May mirrors her Mother's actual spoken voice from the first part of the play, May's interiority is fully effaced: May now has fully and finally become, even as she is a kind of *spectral embodiment of,* the crypt containing—and repeating—the discourse of the Mother: "Amy...Amy...Yes, Mother...Will you never have done? Will you never have done...revolving it all? It?..It all...In your poor mind...It all...It all."[73]

As the play concludes, as May's interiority, compromised by the authoritative presence of the Mother's discourse, falters and fades, we may recall Derrida's reading of the archive: "*there* where men and gods command, *there* where authority, social order are exercised, *in this place* from which order is given."[74] In *Footfall*'s final moments we have the maximum compression of Beckett's complex critique of the archive as ordering principle. The classic view of the archive as that which preserves and conserves history, as in itself maintaining temporality as such insofar as *history* itself is maintained, surely does not hold in this world where histories are interchanged and overwritten according to whatever subjectivity authorizes the narrative of the past. And if, as I have been suggesting, the archive, as classically understood, presents a model of interiority as such—a stable agency orienting itself *to* a clearly defined past *by* a clearly defined past—surely *Footfalls* disrupts the idea of interiority as stability: interiority becomes, to return to my reading of *Not I, the crypt in which the subject is hidden even from itself.* And yet, for all that, May's body, having been effaced, will return (if only in the next performance) to find itself yet again inscribed by the authority of the Mother: May's body—that indivisible remainder—must be present, must be *made* present, in order for the archive's authority to be announced.

Conclusion: The Posthuman Archive

This idea of spectral embodiment, so perfectly and agonizingly represented in May, is the tropological through-line between our three plays. May's question, which ultimately is the absent, now ventriloquized Mother's question, "Will you never have done...revolving it all?" is immediately answerable: no. Like Winnie's objects, now shattered and discarded only to re-appear with the sounding of the bell, May will return as specter to her own life, forever to pace the grounds of her own diminished interiority, forever, like Mouth, to trace the outline of her vanished self. How then does one expiate the *spectral* past, the past which can only ever return to lay claim to the present? How does one comprehend the temporality, and the interiority, of a continually ending subject?

In some ways these questions are unanswerable. More precisely, from a perspective of what we could call a "rational ontology" these questions defy conventional philosophical categories: we are compelled to move beyond ontology into something approaching Derrida's notion of hauntology. In Beckett's terms—as, for instance, outlined in *Proust,* "the subject has died—and perhaps many times—along the way"[75]—we have moved beyond the subject, the subject who himself has moved past ontology, repeatedly. We have, in other words, moved into something approaching the posthuman, certainly the posthumous. Winnie's world is one that comes after some (unidentified and perhaps unidentifiable) cataclysm; Mouth is reduced to a fragment of a (presumably) previous wholeness; May (and Voice) are shades of what they once were. In all sense of the term, from the purely temporal (they come "after" some kind of integrated humanity) to the philosophical (the subject in Beckett radically calls into question all notions of self-coincidental interiority a priori) the Beckettian character is posthuman.

Thus another way of speaking of "embodied spectrality" is to begin thinking—and I have only room here to offer some tentative speculation—of the idea of the *posthuman archive.* Is there not already something implicitly posthuman about the notion of the archive? Is there not something touching on the posthuman in Derrida's notion that the archive responds to a question from the spectral future? And certainly if we begin thinking about the archive as that which stands as testament—as a monument—to what will be lost, what will *have been* lost, the archive itself always already bears within itself the traces of the *after,* the aftereffects of the human, of human culture. Beckett's plays, as I see them, thus are a radicalization of an idea already implicit in the archive: history is contained precisely within the space—be it the embodied spectrality of the posthuman subject; be it within the phenomenal space of the museum or library—that marks the anticipation (perhaps the creation) of history's absolute loss. The archive, in terms I have discussed elsewhere, is always already melancholy, is itself the fetishistic space wherein the loss of history is *continually* staged.[76] It is here that Beckett's specters achieve their most uncanny effect: the stage—as cryptic archive—offers a fragment of the human (*Not I*); the subject as specter (*Footfalls*); the subject as repeating, thus spectral a-temporality (*Happy Days*); but these specters, obsessed as they are with the absolutely recognizable concerns of the human (loss, family, disabling love), speak to their audience: we are confronted with the most radicalized image of a reduced humanity—a posthuman subject— yet we still hear within their voices the appeal to what has been an integrated subject position, an *historical* integrity. This is, perhaps, as much to say that a fully effaced subject is not a possibility for the stage, but I think it is for Beckett more: the residual traces of the past, continually deforming the subject, continually spectralizing the subject, are precisely what constructs her at the limits, construct her as the space between the claims of the past and the appeal of the present. Embodiment thus becomes the means for translating the past, for concretizing the past, making its claims real for the subject as for the audience. The body on stage concretizes the idea that the past is *never* effaced; its claims can never be fully erased. The phenomenal body, as site of the collision between past and present, as the site where that collision speaks the subject into spectral being, thus always functions as an afterimage: *but it is an afterimage always verging into the real.* It is, again and finally Derrida, who offers the most cogent analysis of the truly unsettling effect of the specter:

> There are several times of the specter. It is a proper characteristic of the specter, if there is any, that no one can be sure if by returning it testifies to a living past or to a living future, for the *revenant* may already mark the promised return of the specter of the living being.[77]

We can say then that the posthuman subject in Beckett is the surplus product of the collision of various temporalities of the archive, which in itself is simply to say that the body in Beckett must always itself be compromised, fragmented, divorced from its voice. The (bodily) archive always already must function as an index of its own compromised attempt to maintain, to contain, a history which could, ideally, offer the subject—in the work of mourning—a totalized vision of itself *to itself.* And thus the posthuman subject, finally, is the only viable trope of the archive as it really is: the spectral site of history's effaced continuity, "irremediably part of us, within us, heavy and dangerous."

Notes

1. Samuel Beckett, *A Piece of Monologue,* in *The Collected Shorter Plays of Samuel Beckett* (New York: Grove, 1984), 269.
2. Sigmund Freud, "Mourning and Melancholia," in *The Penguin Freud Library, Volume 2,* ed. Angela Richards (London: Penguin, 1984), 258.
3. See their attempt to "depathologize" melancholia in the introduction to *Loss: The Politics of Mourning,* ed. David L. Eng and David Kazanjian (Berkeley: University of California Press, 2003), 3–4.
4. Beckett, *Proust* (New York: Grove, 1931), 3.
5. *Breath* (1960) is the exception to the rule given that no bodies appear on stage. One might argue after Derrida, however, that the play's insistence on voice—a cry at birth, a cry at death—stages *presence* in its most traditional form.
6. Pierre Nora, "Between Memory and History: *Les Lieux de Memoire,*" *Representations. Special Issue: Memory and Counter-Memory* (Spring 1989): 14.
7. Jacques Derrida, *Archive Fever: A Freudian Impression,* trans. Eric Prenowitz (Chicago: University of Chicago Press, 1995), 1.

8. Ibid., 11.
9. Ibid., 16.
10. Ibid., 36; 84.
11. Ibid., 85.
12. Derrida uses the term "hauntology" in *Specters of Marx* to refer to the specter that does not, he suggests "belong to ontology, to the discourse on the being of beings, or to the essence of life or death." *Specters of Marx: The State of the Debt, the Work of Mourning, and the New International,* trans. Peggy Kamuf (New York: Routledge, 1994), 51.
13. Jacques Derrida, *The Ear of the Other: Otobiography, Transference, Translation,* ed. Christie McDonald (Lincoln: University of Nebraska Press, 1988), 57–58.
14. Dominick LaCapra, drawing on Freud's "Remembering, Repeating, and Working-Through," links melancholia to the concept of "acting out" and mourning to "working through." See *Writing History, Writing Trauma* (Baltimore: Johns Hopkins University Press, 2001), 70.
15. Samuel Beckett, *Happy Days* (New York: Grove, 1961), 7.
16. Jacques Derrida, "Foreword: Fors: The Anglish Words of Nicolas Abraham and Maria Torok," in *The Wolf Man's Magic Word: A Cryptonymy,* trans. Nicholas Rand (Minneapolis: University of Minnesota Press, 1986), xiv.
17. Beckett, *Happy Days,* 42–43.
18. Ibid., 32.
19. The early manuscripts of the play show character B (later Willie) reading newspaper headlines about rocket attacks. S.E. Gontarski, *The Intent of Undoing in Samuel Beckett's Dramatic Texts* (Bloomington: Indiana University Press, 1985), 80. Surely a trace of the idea that this world has been obliterated by military attack lingers in the finished play.
20. Beckett, *Happy Days,* 7.
21. Ibid., 28.
22. Ibid., 8.
23. Ibid., 9.
24. Ibid., 10; 14; 15; 31; 61; 58; 51; 49; 40, 26.
25. Beckett, *Watt,* 79.
26. Ibid., 29.
27. Derrida, *Archive Fever,* 36.
28. Beckett, *Happy Days,* 38.
29. Ibid., 39.
30. Ibid.
31. In some ways the uncanny temporality of Winnie's world should remind us of Freud's conception of the temporality of the unconscious, where "[n]othing can be brought to an end, nothing is past or forgotten." Sigmund Freud, *The Interpretation of Dreams,* in *The Penguin Freud Library, Volume 4,* ed. Angela Richards (London: Penguin, 1991), 577–78.
32. Beckett, *Happy Days,* 50.
33. Ibid., 52.
34. Ibid., 54.

35. Beckett, *Watt*, 72.
36. Samuel Beckett, *Three Novels: Molloy, Malone Dies, The Unnamable* (New York: Grove, 1958), 291.
37. Samuel Beckett, *Not I*, in The *Collected Shorter Plays of Samuel Beckett* (New York: Grove, 1984), 217.
38. Ibid., 216.
39. Ibid., 217.
40. Ibid., 216.
41. See Anna McMullan on the absent Father in *Theatre on Trial: Samuel Beckett's Later Drama* (New York: Routledge, 1993), 81.
42. Beckett, *Not I*, 216.
43. Ibid., 219.
44. Ibid., 220.
45. Ibid., 219.
46. Ibid., 217; 219.
47. Ibid., 217.
48. Ibid., 219–20.
49. Ibid., 219.
50. Ibid., 219.
51. Ibid., 220.
52. We recall Molloy: "Perhaps I'm inventing a little, perhaps embellishing.... But perhaps I'm remembering things" [8–9]; or Moran's more complex "For in describing this day I am once more he who suffered it" [122]).
53. For a very different reading of the valence of the term "imagine," see Ann Wilson, "'Her Lips Moving': The Castrated Voice of *Not I*," in ed. Linda Ben-Zvi, *Women in Beckett: Performance and Critical Perspectives* (Urbana: University of Illinois Press, 1990), 196–97.
54. Beckett, *Not I*, 220.
55. Ibid., 222.
56. Ibid., 220.
57. Ibid., 216; 221; 222.
58. Ibid., 216.
59. Nicholas Abraham and Maria Torok, *The Wolf Man's Magic Word: A Cryptonymy*, trans. Nicholas Rand (Minneapolis: University of Minnesota Press, 1986), 141.
60. Ibid., 142.
61. Sigmund Freud, "Beyond the Pleasure Principle," in *On Metapsychology: The Theory of Psychoanalysis, The Penguin Freud Library, Volume 11*, ed. Angela Richards (London: Penguin, 1991), 283.
62. Samuel Beckett, *Footfalls*, in *The Collected Shorter Plays of Samuel Beckett* (New York: Grove, 1984), 239.
63. Ibid.
64. In an interview with John Connor, Billie Whitelaw related how Beckett, in responding to her queries about whether May is alive or dead, said "Let's just say you're not quite there" (Quoted in Steven Connor, *Samuel Beckett: Repetition, Theory and Text* (Oxford: Blackwell, 1988), 154.

65. Beckett, *Footfalls*, 242.
66. McMullan, in *Theatre on Trial*, notes that the play as a whole stages May's "inability to separate herself from the Mother" (101).
67. Beckett, *Footfalls*, 240.
68. Ibid., 241.
69. Ibid.
70. Derrida, *Ear of the Other*, 58.
71. I am borrowing the resonance of the term "deform" from Beckett's *Proust:* "There is no escape from the hours and the days. Neither from to-morrow nor from yesterday. There is no escape from yesterday because yesterday has deformed us, or been deformed by us" (2).
72. Beckett, *Footfalls*, 242.
73. Ibid., 243.
74. Derrida, *Archive Fever*, 1.
75. Beckett, *Proust*, 3.
76. Jonathan Boulter, "The Melancholy Archive: Jose Saramago's *All the Names*," *Genre* 37.1/2 (Spring/Summer 2005): 115–43.
77. Derrida, *Specters of Marx*, 99.

CHAPTER 8

"…Humanity in Ruins…": The Historical Body in Samuel Beckett's Fiction

Katherine Weiss

Many critics have analyzed Samuel Beckett's fragmented bodies using philosophical and psychoanalytic approaches that, at least when applied to discussions of Beckett, are often detached from historical consideration. Linda Ben-Zvi understands the fragmented body as a sign in her 1992 essay "*Not I:* Through a Tube Starkly," arguing that the mouth in the television version of *Not I* shifts in meaning from that of a mouth to a vagina, both devouring and expelling language simultaneously.[1] Likewise, in her seminal work, *Theatre on Trial: Samuel Beckett's Later Drama,* Anna McMullan dedicates a large portion of her analysis to the fragmented body on the Beckettian stage. She argues that the fragmented body is "a powerfully metonymic device, since it requires of the viewer an imaginary filling-out of the missing body." Beckett's narrowing down of the body, according to McMullan, "changes the status of the visible portion of the body into that of a sign" which the audience feels compelled to fill in through interpreting the unknowable.[2]

My own earlier work, similarly, has examined Beckett's *Not I, That Time* and *What Where* in relation to the way in which viewers read the fragmented body on stage. However, I drew on Gilles Deleuze and Félix Guattari to argue that the works comment on how texts are read and by extension how meaning is constructed through the process of gazing at, what Deleuze and

Guattari term in *A Thousand Plateaus,* the white wall/black hole system of faciality. These white walls and black holes, for the two philosophers, make up the face. Deleuze and Guattari explain that meaning is read through both the words spoken and the automatic, unconscious facial expressions that accompany both the speaker and the one spoken to. The surface of the face becomes a white wall which reflects the appropriate facial expression, and the black holes are the eyes and mouth which absorb meaning. Moreover, the use of masks in primitive rituals, Deleuze and Guattari theorize, is an act to inscribe meaning onto the unknowable. Their prime example of this is the death mask; rituals around death and the employment of the death mask ultimately provide closure to the horror of the unexpressive face of the dead and provide a story that helps those living to cope with the unknown.[3] Hence, what Beckett gives us, as I have argued elsewhere,[4] are masks that expose our own processes of reading images that are unknowable.

Now several years later, I reflect back on my own ahistorical reading of Beckett. I wonder whether the process of reading white walls and black holes in Beckett reflects the way in which those suffering from "historical trauma," trauma resulting from traumatic and unexplainable events involving victimization, (such as when one is faced with mass destruction as Beckett encountered while working with the Irish Red Cross) *act out* and attempt to *work through* traumatic experience. For theorist Cathy Caruth, trauma is "the encounter with death, or the ongoing experience of having survived it." She goes on to explain that "What returns to haunt the victim [...] is not only the reality of the violent event but also the reality of the way that its violence has not yet been fully understood."[5] Drawing on Freud's analysis of *Gerusalemme Liberata* in *Beyond the Pleasure Principle,* Caruth argues that the traumatic event is repressed because of the witness's guilt to surviving the unfathomable encounter with death. The unwelcome repetition of the event is "not just an unconscious act of the infliction of the injury," but rather "a voice that cries out from the wound, a voice that witnesses the truth" that the survivor cannot reconcile.[6]

Like Caruth, Dominick LaCapra theorizes that survivors of catastrophic historical events are "possessed by the past and performatively caught up in the compulsive repetition of traumatic scenes—scenes in which the past returns and the future is blocked or fatalistically caught up in a melancholic feedback loop." In, what LaCapra calls the *acting out* of the event, that is an unwelcome return, "tenses implode, and it is as if one were back there in the past reliving the traumatic scene. [...] In this sense the aporia and the double bind might be seen as marking a trauma that has not been worked through."[7] "Acting out," for LaCapra who is indebted to Freud, "is a process but a repetitive one."[8] It comes back, repeatedly, haunting those affected by traumatic events and keeping them from obtaining the critical distance needed to examine the experience and the event. While obtaining this distance and being able to construct a controlled narrative of the event, a process LaCapra calls *working through* trauma, does not release one from the ghosts of the past (it is not, in other words, a cure), it does allow one to come to terms with traumatic events: "In working through, the person tries to gain critical distance on a problem and to distinguish between past, present, and future."[9] LaCapra warns that *acting out* and *working through* are not binary oppositions, but rather similar processes both of which involve repetition, and both of which may coexist in a person. However, the repetition involved in the process of *working through* is not one that is "compulsive"; it is, instead, a process that leads "to a rethinking of historicity and temporality in terms of various modes of repetition with change."[10] To bring the matter back to Beckett, LaCapra argues that much of postmodern experimental literature is a controlled response to trauma—a process of *working through* traumatic experience. Perhaps, as Caruth points out, this is so because "literature, like psychoanalysis, is interested in the complex relation between knowing and not knowing."[11]

In this chapter, I move away from a study of Beckett's drama to examine three of his experimental short prose works from the 1960s, *All Strange Away* (1963–1964), *Ping* (1966) and *Lessness* (1969), suggesting that even twenty years after his work with the Irish Red Cross, Beckett was still haunted by the experience of arriving in Saint-Lô where he witnessed the result of World War II's mass destruction and his guilt of being a national of Ireland—one of the few nations that declared itself neutral during World War II. Theorist Dominick LaCapra and psychoanalyst Adam Phillips have both written about the forms which trauma takes in modern art and writing. Phillips draws attention to three aspects of modern art, all of which Beckett's fiction of the 1960s displays: repetition, the exact positioning of things, and distortion, often experienced in Beckett as fragmentation. Drawing on Freud, Phillips, like LaCapra and Caruth, argues that "Repetition is the sign of trauma; our reiterations, our mannerisms, link us to our losses, to our buried conflicts,"[12] and he continues to point out that writing is "a form of repetition that easily obscures its own history, the conflicts it was born out of, the problems which made it feel like a solution."[13] Unlike LaCapra who sees writing as an attempt to *work through* trauma, Phillips perceives writing as testifying to both processes—*acting out* and *working through* the experience. Not only is Beckett's writing rife with repetition (one needs only to think of Vivian Mercier's famous 1956 review of *Waiting for Godot,* in which he claimed that Beckett had done the impossible, he had written a successful play "in which nothing happens, *twice,*"[14] but also the very act of writing, the struggle to *work through* the past, is a re-seeing, or re-saying that is sometimes itself distorted by a lack of distance; the writing of the past, in other words, is often *ill seen* and *ill said*.[15]

The second indicator of internal conflict, or trauma, according to Phillips, is "the need for things to have an exact position—the determined commitment to a definitive historical narrative."[16] While Beckett's texts are not historical narratives, that is they are not attempts to reconstruct or revise historical events,[17] there is in some of Beckett's texts, especially *Endgame* and his short fiction of the 1960s, an attempt to situate the body in an exact position. The blind and wheel-chair bound Hamm absurdly tries to ensure that his servant Clov has returned him to the exact center after taking Hamm on a little tour of the stage. Similarly, in the fiction of the 1960s, the narrator geometrically assigns positions to the body. We read in *All Strange Away*: "Call floor angles deasil a, b, c and d and ceiling likewise e, f, g and h, say Jolly at b and Draeger at d, lean him for rest with feet at a and head at g, in dark and light, eyes glaring, murmuring."[18] This desire to position the body in an exact location reflects, in some sense, the historians need for exactness. Yet as LaCapra discusses in *Writing History, Writing Trauma,* the historian's need for accuracy is thwarted by the impossibility for the survivor or witness to provide an exact account of traumatic events. He, nonetheless, urges that their "inexact" accounts of traumatic events make the experience no less accurate, or true. Such testimony bears witness to the intimacy of trauma of the individual survivor.

Beckett's fragmented bodies reflect Phillips' third and main argument, concerning the distorted body of artists (Picasso being his primary example). Phillips explains that trauma may be defined as a "too-closeness," or the lack of critical distance. Survivors have no distance from the traumatic event, he argues; their experience is one of intimacy and as such is manifested in distortion and fragmentation.[19] It is this last element of Phillips' argument that is perhaps the most important to this study of Beckett's prose fiction of the 1960s. The "too-closeness" Beckett experienced when confronted with the mass destruction caused by World War II haunted him throughout his writing. Even when he did not write about ruins or the human condition under catastrophic duress as he does in *The Unnamable, Waiting for Godot, Happy Days* or the fiction of the 1960s, he dealt with men who were haunted because of a personal catastrophe, such as in *Eh Joe.* Joe *acts out* the catastrophic suicide of one of his former lovers until his face, for the viewer, is distorted by the camera's extreme close-up. Beckett, hence, suggests that despite this man's attempt to suppress the memory of the suicide, he is haunted by the past, acting it out, and his too-closeness to the event, or trauma, is reflected in the close-up of the camera. The camera, inching its way forward throughout the piece, comes in frightening close, distorting Joe's face.

Those that overlook the relevance of historical trauma in Beckett's texts, moreover, have ignored Beckett's own role in political history. Beckett was more than mindful of the political injustices that occurred during his lifetime; he worked with the French Resistance during World War II and thereafter joined the Irish Red Cross. He wrote plays and novels, furthermore, that ridiculed tyrants. Pozzo in *Waiting for Godot* (1948), the policeman in *Molloy* (1955) and the Director in *Catastrophe* (1982) testify to this. His short and somewhat awkward play *Catastrophe* was written for AIDA (the International Association for the Defense of Artists) to honor the Czech dissident writer Vaclav Havel who had been under house arrest for several years and then later imprisoned.[20] In fact, in Beckett studies, scholars either recount Beckett's World War II movements as biographically important (his past involvement helps us to better understand Beckett, the man, in relation to his work) or the discussion of his role in these historical events is absent all together.[21] Although my study will repeat some of the more famous narratives about Beckett's involvement during World War II and Beckett's work in the reconstruction of Saint-Lô from which he wrote "The Capital of the Ruins," a commentary for Radio Eireann, I hope not to give a purely biographical reading of his later prose fiction. Rather I wish to argue that Beckett's fragmented textual style and the fragmented body in *All Strange Away, Ping,* and *Lessness* may be conceived as works which can be read as a representation of the process of *acting out* trauma and Beckett's own *working through* of it. His narrators and characters often suffer from being too close to the traumatic events; however, I suggest that in Beckett the too-closeness of the narrators and characters works to guide the readers into critical distance (as was Beckett's contemporary Bertolt Brecht's goal with the *Verfremdungs Effekt*).

Critics, such as Richard Gilman,[22] desire to read Beckett as ahistorical. However, the very movements that scholars such as Martin Esslin[23] align Beckett with (Existentialism and Theatre of the Absurd) as well as themes that scholars have continually discussed in relation to Beckett, mainly "alienation, absurdity and meaninglessness," are, as Marjorie Perloff rightly points out, directly related to a post–World War II experience.[24] Out of the ashes of World War II, these theories challenged Hegelian notions of progress and existence.

Walter Benjamin, the German philosopher who would later die fleeing Nazi persecution, saw a shift in the narratives that emerged out of World War I trauma. He argued that storytelling and literature no longer served to counsel oneself and others; they were no longer, in other words, communal acts. "Was it not noticeable at the end of the war," Benjamin asks, "that men returned from the battlefield grown silent—not richer but poorer in communicable experience?"[25] Elaine Scarry's 1985 book *The Body in Pain: The Making and Unmaking of the World* eerily echoes Walter Benjamin's claims about the ways in which modes of expression changed as a consequence of mass destruction. Scarry argues that pain destroys language. "Physical pain," she writes, "does not simply resist language, but actively destroys it."[26] The individual in pain is reduced "to the sounds and cries a human being makes before language learning."[27] She draws a parallel between the destruction of a city and of the human body, arguing that both are a destruction of civilization and the ultimate destruction of civilization is to destroy its language.[28] Hence, according to Scarry, war, in its goal to out-injure its opponent, also has as its goal to destroy the other's language and tools of communication. Codes are, then, an attempt not only to communicate to one's side vital information, but also a way to hide language and communication from the opposition to keep it from being destroyed. What is intriguing is that, while each side attempts to destroy the other's language, both sides, as Benjamin suggests, suffer from a break-down in communal language. Like bombed out buildings, the powers of communication, too, must be reconstructed after war.

I venture to argue that this destruction of language and of storytelling of which Scarry and Benjamin theorize is closely linked to historical trauma. Unable to comprehend and as such unable to express the mass destruction of a landscape, cityscape or of the body, writing logical, clear, and aesthetically beautiful texts become causalities of violence and war. Indeed, if writing "poetry after Auschwitz is barbaric,"[29] then the voice of poetry written after the Holocaust must be a barbaric, primitive yelp. It is not until the repressed trauma haunts the victim that language can begin to reconstruct and *work through* traumatic experience.

Marjorie Perloff is one of the few scholars today who has examined Beckett's prose fiction in relation to his war experience. When Beckett's cell of the Resistance was discovered in 1942, he and his partner Suzanne Deschevaux-Dumesnil underwent a "hazardous and painful" journey of roughly 700 km on foot in which he experienced feelings of fear and boredom in order to flee the Gestapo.[30] Perloff argues that in three of his short stories, "The Expelled," "The Calmative," and "The End," written in 1946, Beckett draws on the dangerous journey to Roussillon. By rooting the seemingly paranoid and irrational narrators of these three tales in history, according to Perloff, they no longer appear to be strange. Although the narrators still display paranoia, their behavior makes sense as they are refugees who desperately seek shelter and fear capture. Yet Beckett does not provide us with traditional historical fiction—what the narrators flee (and the flight itself) is obscured either because Beckett lacked critical distance from the traumatic events he had only just experienced, or because having critically distanced himself he can write about the lack of critical distance. Like the texts I am dealing with, "The Expelled," "The Calmative," and "The End" reveal historical trauma, and depict a way to write about and as such *work through* the fragmented civilization of post–World War II Europe.

While Beckett's protagonists in his pre-war fictions are far from "normal," their bodies are intact. Indeed, not until *Watt* does Beckett infer that the body may be distorted.[31] Although Watt's body is not fragmented, detailed accounts about, for example, the way in which he walks raises doubt as to the normality of his body.[32] Beckett's protagonists in the *Trilogy* (1955–1958) "decompose"; they are and become increasingly crippled and ultimately are reduced to crawling. In his later prose fiction such as *How it is* (1961), the body is further reduced, mirroring the reduction of Beckett's language and the textual space is one of a mudscape. In works such as *All Strange Away, Ping,* and *Lessness,* the body appears only as fragments. They no longer constitute a speaking subject. Reflecting the fragmentation of the body, the prose of this time is composed primarily out of sentences which often lack articles, prepositions, verbs and/or subjects. Whereas the fragmented body on Beckett's stage (for example, *Not I* and *That Time*) challenges the conventions of theatre by trying to show the internal trauma of his characters (which can be read in relation to theatre history as well as psychoanalysis), the significance of the fragmented body and text in Beckett's prose fiction stems, I argue, from his war experience.

After the war, Beckett moves further away from the whole body, poking holes in his characters such as Molloy and Moran who lose the ability to walk, and the Unnamable who is a speaking subject without a body. According to Scarry, when we try to describe the body in pain, we "conceptualize a weapon or object inside the body."[33] And, indeed, in *How it is* the body in pain, the body surrounded by mud, on a couple of occasions is attacked by some other. On one such occasion the narrator of *How it is* jabs another being by the name of Pim in the armpit with his nails so that Pim would resume his song. Years earlier, in *Molloy,* the protagonist knocks his mother on the head to communicate to her his desire for money. This act of communication like the jabbing in *How it is* resembles acts of torture. Both post–World War II novels reveal that these acts of violence are done to extract information or materials. While acts of torture in *Molloy* and *How it is* suggest a haunting of the historical and political events of World War II, in these novels Beckett distances the acts of torture from a historical or political reading, unlike his play of 1983, *What Where.* Ultimately, however,

Molloy's mother does not understand what her son wants from her, and thus like the acts of torture carried out to discover what and where in *What Where,* the goal is never achieved.

Prior to World War II, Beckett's short stories and novels were perhaps too rigidly influenced by another great Irish writer, James Joyce. However, while hiding from the Gestapo after his cell of the Resistance was exposed, Beckett takes on a new form of writing—one of fragmentation. Beckett described this new aesthetic as one of "impotence, ignorance" in an interview with Israel Shenker in 1956.[34] His short fiction of the 1960s are like musical tunes that falter, yet do so appropriately, expressing this impotence and ignorance. Yet Beckett's decision to write about impotence and ignorance was not merely a way to distance himself from Joyce. As his writerly goal "to fail, as no other dare fail"[35] does not take full form until after World War II, we can speculate that Beckett's works of fragmentation and failure are rooted, at least in part, in his experience in Saint-Lô, Normandy as a quartermaster, interpreter and driver for the Irish Red Cross in 1945. While, according to Knowlson, Beckett joined the Irish Red Cross solely "as a means of getting back to France and keeping his apartment legally,"[36] he was unprepared for the destruction he was confronted with, and this destruction would haunt his writing of the subsequent years.

Before I dive into that fertile muck of Beckett's writing from the 1960s, I will examine Beckett's own non-fictional account of his postwar activities. In 1945 Beckett was desperate to return to his beloved France. To aid his return, as already stated, he joined the Irish Red Cross and found himself working in the Normandy town of Saint-Lô, where he saw the mass destruction and devastation caused by the war. Although grateful for the chance to return to France after the war, being confronted with the extent of the destruction in Saint-Lô was undeniably a shocking experience. He once again witnessed individuals who had to survive under extreme circumstances and once again he had to come to terms with the destruction he was fortunate enough to have escaped. Along with not having experienced the massive destruction of the war first hand, Beckett's guilt was exasperated by his country's neutrality.[37] Beckett was so stunned that in 1945, he wrote to his close friend Thomas MacGreevy: "St.[-]Lô is just a heap of rubble, la Capitale des Ruines as they call it in France. Of 2600 buildings 2000 completely wiped out, 400 badly damaged and 200 'only' slightly."[38] Of his experience, Beckett in 1946 wrote a short piece for Radio Eireann, called "The Capital of the Ruins," which as Phyllis Gaffney points out, in her informative essay of 1999 "Dante, Manzoni, De Valera, Beckett...? Circumlocutions of a Storekeeper: Beckett and Saint-Lô," was probably never broadcast because of its rather harsh look at Ireland's ineffectiveness to help restore Saint-Lô and in contrast its respectful nod at the inhabitants of Saint-Lô for their will to go on.[39]

Despite countless efforts by students and scholars to view Beckett as the ultimate pessimist, "The Capital of the Ruins" is oddly uplifting in its recognition and celebration of survival. The "heroic period" which Beckett records is not the so-called generosity of the Irish Red Cross nor his native Ireland whose neutrality during the war left Beckett with serious guilt and doubts as to Ireland's long fought for independence.[40] Rather Beckett commemorates the citizens of this Normandy town whose "smile at the human condition" cannot be "extinguished by bombs" or "broadened by the elixirs of Burroughes and Welcome,—the smile deriding, among other things, the having and the not having, the giving and the taking, sickness and health."[41] Beckett sees these citizens as resilient; they are heroic in their refusal to give up. Their will to survive is constant; it does not diminish when faced with their horror of rubble, malnutrition, sickness or injuries caused by masonry falling on them or children stepping on mines, nor does it broaden when presented with the Irish Red Cross's alms. He goes on to describe how, in this town that was

> [...] bombed out of existence in one night, its population of German prisoners of war, and casual labourers attracted by the relative food-plenty, but soon discouraged by housing conditions, continue, two years after the liberation, to clear away the debris, literally by hand.[42]

By including the German prisoners of war in his account of the reconstruction of Saint-Lô, Beckett reveals his own true generosity and forgiveness. He avoids scolding those that during the catastrophic conflict were the enemy. Rather he shakes his finger at the Irish Red Cross's naiveté. He correctly does not, as his fellow Irishmen and women do, believe that it will take a mere ten years to rebuild Saint-Lô; it took "the best part of two decades" for the restoration in Saint-Lô to be completed.[43] For Beckett this image of clearing away the rubble by hand in order to rebuild the town is "a vision and sense of a time-honoured conception of humanity in ruins."[44] In this radio piece, Beckett offers the Irish public a glimpse at a population that refuses to allow the ruins of their town destroy their humanity. Interestingly, it is this image of humanity that haunts Beckett as much as the catastrophic events that led to the destruction of Saint-Lô. Like the Protagonist in *Catastrophe*, who has been subjected to humiliation and torture so that the Director may orchestrate a catastrophic event on stage, these citizens and prisoners of war raise their heads, even if only slightly, to defy and defeat the tyranny that has left them with nothing.[45] No matter how destitute their situation, they continue to work towards restoration.

The "conception of humanity in ruins" is key to understanding Beckett's fragmented textual bodies of the 1960s. The narrator of *How it is,* for example, is not drowning in the mud of his "ruinstrewn land" as it is called in *Fizzle 3*.[46] He crawls through the mud, which Ruby Cohn, H. Porter Abbot and others have argued is an image of fertilization and creation, attempting to build a narrative.[47] The mud, however, is not merely a biblical reference to creation; it is a very real reference to the mud and slime haunting Beckett that made working and rebuilding towns like Saint-Lô almost impossible.[48] When the rain came, the ruined landscapes became mudscapes. Mud, in fact, hinders construction work as houses and buildings must be built on solid foundations. But this knowledge in no way disputes the claim that the narrator in *How it is* creates out of the muck he crawls through. Rather Beckett's novel testifies to the human spirit once again; despite the seemingly impossible task of expelling the mud from his mouth to narrate how it is, the narrator, in fact, is capable of building a story of struggle and survival. Indeed, there is a shift in Beckett's prose of the 1940s and 1950s of narratives that depict the body falling apart to the narratives of the 1960s in which the storyteller struggles to piece the body together. The prose fiction that follows *How it is* may be one of extreme fragmentation, but the narrators of these texts strive to build, or rebuild civilization out of ruins.

Much of modern fiction, as LaCapra, Caruth, and Phillips posit and Beckett's texts reveal, attests to historical trauma. Such fiction is like the heaps of rubble, a sign of the individual's inability to communicate, to connect with and empower the individual, and yet in Beckett's fiction there exists a continual attempt to connect and restore communication. Beckett suggests, however, that while we can attempt to restore the bombed out cities, evidence of the destruction, and by extension evidence of trauma even after *working through* trauma, will always exist; we will always be haunted, Beckett seems to whisper to his reader, by the catastrophic events of the past. About the impossibility to erase the presence of the Irish in Saint-Lô, he writes in "The Capital of the Ruins": "But I think that to the end of its hospital days it will be called the Irish hospital, and after that the huts, when they have been turned into dwellings, the Irish huts."[49] Although Beckett strives to restore the destruction of communal language and storytelling, he discovers that it is impossible to wipe out evidence of the traumatic destruction of place and language.[50] In Beckett's fiction, the fragmented storyteller attempts to rummage through and clear away the debris to compose a new textual body out of the rubble, yet this act does not hide the traumatic past, and as such the act of restoration is never complete. The narrator is always haunted (*acting out*) and trying to understand (*working through*) his trauma.

Written between 1963–1964, *All Strange Away* begins with the much written about phrase: "Imagination dead imagine."[51] These opening words ask the reader to imagine a time when imagination will no longer exist, when the unspoken language of one's mind has been obliterated. What Beckett's work suggests, however, is the impossibility of this erasure because the very act of imagining the death of imagination is an act of "fancy." In the OED, as Darren Gribben points out, "fancy" is synonymous with "imagination." Hence, he rightly argues that the capacity to imagine makes the imagined real.[52] Despite being asked to contemplate this state, the reader cannot because in the mere contemplation of the death of imagination, imagination flourishes.

This attempt to imagine the death of imagination coupled with a brief reference to war is curious. In *All Strange Away* Beckett writes: "Out of the door and down the road in the old hat and coat *like after the war,* no, not that again."[53] Here, the narrator begins to tell a tale of temporary or permanent expulsion after the war. Is this, the reader ponders, what we are not to imagine? With the phrase, "no, not that again," Beckett both denies the reader a postwar narrative (no, I will not tell that story again) and reveals that the postwar narrative has already been said (again). He, in other words, tries to wipe out the memory of war and life after the war from his imagination but in doing so refers to the war, and is perhaps haunted by it. The narrator attempts to encrypt his war experience like the layers of mud that cover the ruins of a bombed out town after the rain. These encrypted stories not only hide "the real story" but also work to form a new type of storytelling. The narrator both *acts out* the unwelcome memories and strives to *work through* his trauma by bringing the pieces together to understand them and form perceivable images. What he begins to construct are body parts being fixed in a barren and ruined space—body parts that will eventually form an image in the reader's consciousness (much like the fragmented body in Beckett's *Not I* and *That Time*) which ask the viewer to piece together the narrative heard and the body partially seen. Through the fragmented memories and images offered up in Beckett's text, the reader gains a glimpse at a larger picture, but that image is never fully imagined as it would have been. Yet this act of construction, too, is a post–World War II tale—one that reveals the destruction of storytelling and the struggle to reconstruct it.

By the time we arrive at the short fiction of the 1960s, the body has been literally blown to bits. Elaine Scarry argues that the verbal strategy to pain is twofold: the body is conceived as being injured by a weapon or object and as being fragmented. Phillips, too, points to the destruction of the body when he argues that artistic renditions of the body as distorted is the result of a "too-closeness" with traumatic experience. In *All Strange Away* and

Ping, body parts litter the textual space. After the "narrative voice" casts off his memories of the war and its aftermath in *All Strange Away,* it asks the reader to imagine a "last person, murmuring."[54] The way in which the voice describes this last person and another body, that of a female, is through a series of body parts positioned in various places in the text. Several times, we are told to imagine

> Arse to knees, say bd, feet say at c, head on right cheek at a. Then arse to knees say again ac, but feet at b and head on left cheek at d. Then arse to knees say again bd, but feet at a and head on right cheek at c.[55]

This positioning is an extreme attempt at exactness. The narrator, striving for accuracy, literally graphs the body parts onto the page. In these strange, fragmentary images of the body and even stranger narrative almost resembling a graph or diagram of some sort, we can imagine the body moving. The feet move from c to b to a; the arse from bd to ac back to bd; and the head moves from right to left and back to right. This action goes on throughout the text.

Likewise, in *Ping,* Beckett's very short prose work of 1966, the body appears, or just appears, as fragments. In this short work, the first sentence reads: "All known all white bare white body fixed one yard legs joined like sewn"[56] and the narrative continues to fix body parts such as palms, heels, eyes, and lashes onto the geographic space of the page. This all white body is nearly invisible; it leaves only "Traces blurs light grey almost white on white."[57] The faintness of these body parts, slowly appearing as only traces of a body, recalls a ghostly presence, haunting the narrator. This narrative, then, attempts to reveal the way in which victims of trauma *act out* the experience while simultaneously constituting a *working through* it. Suffering from some traumatic experience that the reader is not privy to, the narrator recalls traces of a memory of that traumatic event. But like the reader, he is unable to come to terms with it; he is left trying to read the unreadable.

Despite the aggressive fragmentation in both *Ping* and *All Strange Away,* unlike *The Trilogy,* the body is not disintegrating. Rather there is a move to reconstruct these ruined bodies. In *Ping* this move is apparent in the word "given"; the all white body, almost invisible, is "given rose" and "given blue" and there are "Traces blurs signs" of life even if that life has "no meaning." More explicitly in *All Strange Away,* the two bodies take part in an act of copulation: "Imagine him kissing, caressing, licking, sucking, fucking and buggering all this stuff, no sound"[58] and "Fancy her being all kissed, licked, sucked, fucked and so on by all that, no sound, hands on knees to hold herself together."[59] John Pilling, Peter Murphy, and Graham Fraser have used the term "pornographic" to describe *All Strange Away*.[60] However pornographic, these images also on some level represent acts of regeneration and reproduction. These bodies, or the imagining of these bodies, no matter how destitute, continue to try to rebuild, whether it is a family or bonds of physical intimacy. Yet, the cold and calculating description of these sexual acts and the attempts to pin down the bodies suggest that these bodies will not reach their intended goal of release. They are stuck, repeating compulsively their acts, unable to exist in the present and unable to progress into the future.

Like *Ping* and *All Strange Away,* in *Lessness* the body once again appears as fragments. In this work from 1969, Beckett begins with the haunting image of "Ruins true refuge."[61] Here, the ruins are the landscape that has been destroyed, the body that has been fragmented and the rhythmic poetry of the text. The opening line and others such as "Blacked out fallen open four walls over backwards true refuge issueless" and "Scattered ruins same grey as the sand ash grey true refuge," which are frequently repeated in the five pages that make up this story, both define the ruined space as one devastated by some violent force, perhaps the bombs that scarred Europe, and as the only true protective shelter. Some twenty years earlier, Beckett describes the ruined refuge more clearly in "The End": "The door had been removed, for firewood, or for some other purpose. The glass had disappeared from the windows. The roof had fallen in at several places."[62] Perloff compares the narrator's shed in "The End" with that of the ruined town of Saint-Lô.[63] On the one hand, *Lessness* gives us an even bleaker image of World War II and its aftermath than "The End" does; an image that reveals that individuals had no true refuge from the elements. They had to rummage through the fallen walls of what were once their homes and the grey deluge of what were once their streets and fields. The lucky ones lived in the cellars beneath the rubble.[64] The refuge is one that exposes the narrator to the harshness of the natural elements and to possible injury and threat. However, it is also a refuge that opens the walls of the mind and allows for the process of *working through* traumatic memory.

On the other hand, this image of "humanity in ruins" is one that reveals that even in the most devastating circumstances sanctuary and hope exists. In *Lessness,* like in most of Beckett's work up to this point in his career, yet another body does not give up. This ruined body's heart beats; he breathes: "He will live again the space of a step it will be day and night again over him the endlessness."[65] And five times, we are told that "he will make it," echoing Beckett's radio play *Cascando* of 1962, in which the protagonist Woburn, despite falling down continually, gets up and continues on his journey.

Moreover, in *Lessness* there are hauntings of the past which will come again: "On him will rain again as in the blessed days of blue the passing cloud"[66] and "Old love new love as in the blessed days unhappiness will reign again."[67] The "blessed days" of blue skies, passing clouds and both old and new love will return. The past, however, is not idealized as a time of happiness and bliss. Rather we are told that it will "rain on him" and that "unhappiness will reign again." The pun on "precipitation" and "rule" reveal that the past was not one of sunny liberation. With devastation comes a new freedom, however, and as such a new sense of security. Indeed, the narrative voice suggests that artistic freedom previously was perhaps only imagined: "Never but imagined the blue in a wild imagining the blue celeste of poesy."[68] Beckett does not argue that the act of creating sets us free; there is no "freedom" gained in the artistic expression as there was no freedom for those who reconstructed Saint-Lô. Instead, as he voiced in "Three Dialogues" the artist has an obligation to express even when there was nothing to express and nothing with which to express.[69] Furthermore, art, for Beckett, is an act of failure as it painstakingly emerges out of unhappy and, at times, inhuman social and political structures. It emerges when haunted by catastrophic events.

Samuel Beckett's experimental prose of the 1960s is an attempt to reimagine poetry and fiction after World War II. For Beckett, there was no way to return to narratives of "omniscience and omnipotence," not after the destruction of the war. Taking the "scattered ruins"[70] lodged in the imagination, Beckett creates a new form of writing, one that does not ignore or silence the destruction and trauma of the war experience. And, in doing so, he ultimately proves Walter Benjamin and Elaine Scarry wrong by creating a remarkably imaginative language out of the ruins of fragmented culture and traumatic experience. He creates distorted and fragmented images, but images that attest to the importance of gaining critical distance and rebuilding civilizations.

Notes

1. Linda Ben-Zvi, "*Not I:* Through a Tube Starkly," in *Samuel Beckett*, ed. Jennifer Birkett and Kate Ince (London and New York: Longman, 2000), 259–65.

2. Anna McMullan, *Theatre on Trial: Samuel Beckett's Later Drama* (London: Routledge, 1993), 161.

3. Gilles Deleuze and Félix Guattari, *A Thousand Plateaus: Capitalism and Schizophrenia*, trans. Brian Massumi (London: Athlone Press, 1999), 167–80, 186–90.

4. See Katherine Weiss, "Bits and Pieces: The Fragmented Body in *Not I* and *That Time*," in *Other Becketts*, ed. Daniela Caselli, Steven Connor and Laura Salisbury (Tallahassee: Journal of Beckett Studies Books, 2002), 187–95.

5. Cathy Caruth, *Unclaimed Experience: Trauma, Narrative, and History* (Baltimore and London: Johns Hopkins University Press, 1996), 2–3.

6. Ibid.
7. Dominick LaCapra, *Writing History, Writing Trauma* (Baltimore and London: Johns Hopkins University Press, 2001), 21.
8. Ibid., 148.
9. Ibid., 143.
10. Ibid., 149.
11. Caruth, 3.
12. Adam Phillips, "Close-Ups," *History Workshop Journal* 57 (2004): 142.
13. Ibid.
14. Vivian Mercier, "The Uneventful Event," in *Critical Essays on Samuel Beckett: Critical Thought Series 4,* ed. Lance St. John Butler (Aldershot: Scholar Press, 1993), 29.
15. Beckett's late novella of 1981, which encompasses the problems surrounding remembering and forgetting, bears the title *Ill Seen Ill Said.*
16. Phillips, "Close-Ups," 146.
17. S.E. Gonatarski argues in *The Intent of Undoing* (Bloomington: Indiana University Press, 1985) that with each revision of his texts, Beckett erased biographical elements from his texts. James McNaughton's essay in this collection, in part, carries this line of reasoning further; however, he makes the case that the residue Beckett left behind urges his readers to read his novel *Watt* biographically and historically, even when at times they are left without clarity. Beckett does this, McNaughton speculates, to point out the workings of propaganda.
18. Samuel Beckett, *All Strange Away,* in *The Complete Short Prose 1929–1989,* ed. and with an introduction and notes by S.E. Gontarski (New York: Grove Press, 1995), 171.
19. Phillips, "Close-Ups," 147–48.
20. James Knowlson, *Damned to Fame: The Life of Samuel Beckett* (New York: Touchstone, 1997), 595.
21. With the exception of Marius Buning et al., eds., "Historicising Beckett/Issues of Performance," *Samuel Beckett Today/Aujourd'hui* 15 (Amsterdam: Rodopi, 2006) and the contributors of this collection.
22. See Richard Gilman, "Beckett," *Partisan Review* 41 (1974): 56–76.
23. The Theatre of the Absurd and its extreme form, Existentialism, are movements that were "characterized by a politicized sense of futility and meaninglessness in a world constantly under the threat of war and nuclear annihilation." See Matthew Roudané, *Dramatic Essentials* (Boston and New York: Houghton Mifflin, 2009), 12. See also Martin Esslin, *The Theatre of the Absurd,* 3rd ed. (London: Penguin, 1980).
24. Marjorie Perloff, "'In Love with Hiding': Samuel Beckett's War," *Iowa Review* 35.2 (2005): 78.
25. Walter Benjamin, "The Storyteller," in *Illuminations,* ed. and trans. Hannah Arendt (New York: Schocken, 1968), 84.
26. Elaine Scarry, *The Body in Pain: The Making and Unmaking of the World* (New York and Oxford: Oxford University Press, 1985), 4.

27. Ibid.
28. Ibid., 61.
29. Theodor W. Adorno, "Cultural Criticism and Society," in *Prisms,* trans. Samuel Weber and Sherry Weber (Cambridge, MA: MIT Press, 1981), 34.
30. Perloff, "In Love with Hiding," 81.
31. In his essay for this collection, McNaughton reads *Watt* as a war novel that erases the war in order to expose and criticize the form in which propagandistic writing takes. In addition, I would suggest that *Watt* displays a too-closeness to the war and the politics of Nazi Germany. As such its writing of the war becomes distorted to the point of erasure.
32. In 1962, the late Hugh Kenner in his book *Flaubert, Joyce and Beckett: The Stoic Comedians* pointed out that the inventories Beckett goes through in his fiction (Watt's way of walking, Molloy's sucking stones) are, in effect, acts of creation—not making something new in the Poundian sense, but instead remaking out of the existing elements in a closed field. Yet, again in Beckett, this too-closeness to the traumatic event results in a distortion of the body. See Hugh Kenner, *Flaubert, Joyce and Beckett: The Stoic Comedians* (London: Dalkey Archive Press, 2005), 67–107.
33. Scarry, *The Body in Pain,* 17.
34. In a comparison with his writing to James Joyce's, he told Shenker that Joyce was "tending towards omniscience and omnipotence as an artist. I am working with impotence, ignorance." Quoted in James Acheson, *Samuel Beckett's Artistic Theory and Practice: Criticism, Drama and Early Fiction* (Basingstoke: Macmillan, 1997), 6.
35. Samuel Beckett, *Disjecta: Miscellaneous Writings and a Dramatic Fragment,* ed. Ruby Cohn (London: John Calder, 1983), 145.
36. Knowlson, *Damned to Fame,* 313.
37. While Beckett never directly condemned Ireland's neutrality, the risks he took in the Resistance testify to his own inability to remain silent in the face of Nazi terror. For more on Beckett's role in the Resistance, see Knowlson, *Damned to Fame,* 273–90. Even after the war, Beckett noted in an interview with Shenker that he "preferred France in war to Ireland at peace" (qtd. in Anthony Cronin, *Samuel Beckett: The Last Modernist* (New York: HarperCollins, 1997), 310.
38. Knowlson, *Damned to Fame,* 313.
39. Phyllis Gaffney, "Dante, Manzoni, De Valera, Beckett...? Circumlocutions of a Storykeeper: Beckett and Saint-Lô," *Irish University Review* 29.2 (Autumn/ Winter 1999): 260.
40. Beckett, who had lost friends to the atrocities of the Nazis, was appalled by Eire's decision to remain neutral, a decision that Elizabeth Bowen claims was overwhelmingly the desire of the people of Eire. The public's support for Ireland's neutrality, Bowen however points out, was tainted by Ireland's censorship laws. Although there was freedom of speech, there was "no freedom of reporting. No home criticism of Eire's neutrality, or suggestion that this ever could or should be abandoned, [was] allowed mention by the press." See *The Mulberry Tree: Writings of Elizabeth Bowen* (London: Harcourt Brace Jovanovich, 1986), 32.

41. Samuel Beckett, "The Capital of the Ruins," in *The Complete Short Prose,* 277.

42. Ibid.

43. Phyllis Gaffney, *Healing Amid the Ruins: The Irish Hospital at Saint-Lô* (Dublin: A&A Farmar, 1999), 49.

44. Beckett, "The Capital of the Ruins," 278.

45. In *Damned to Fame,* Knowlson recalls that Beckett explained to a reviewer that the conclusion of *Catastrophe* in which the Protagonist raises his head slightly was an act of defiance. Angrily, Beckett told the reviewer, "He's saying: You bastards, you haven't finished me yet!" See Knowlson, 597.

46. Samuel Beckett, *Fizzle 3,* in *The Complete Short Prose,* 232.

47. It has been pointed out repeatedly that the French *Comment c'est* is a pun on *commencer* (to begin). See Ruby Cohn, *Back to Beckett* (Princeton: Princeton University Press, 1973), 230; H. Porter Abbott, "Beginning Again: The Post-narrative Art of *Texts for nothing* and *How it is,*" in *The Cambridge Companion to Beckett,* ed. John Pilling (Cambridge: Cambridge University Press, 1994), 111. With no adequate English equivalent, Beckett settled for *How it is* perhaps tempting his readers to fill in "to begin."

48. In Gaffney's history of the Irish Red Cross in Saint-Lô, she quotes from a telling letter which Beckett wrote soon after his arrival: "It has been raining hard the last few days and the place is a sea of mud. What it will be like in winter is hard to imagine. No lodging of course of any kind." See Gaffney, *Healing Amid the Ruins,* 28.

49. Beckett, "The Capital of the Ruins," 278.

50. Beckett would return to Saint-Lô in his writing, Gaffney argues, in *Waiting for Godot, Endgame,* and his poetry (as in "Mort de AD"). Moreover, the presence of the Irish Red Cross, to this day, has never left those living in Saint-Lô, evident among other ways in the fact that one remaining hut serves as a school. See Gaffney, *Healing Amid the Ruins,* 72–77.

51. Beckett, *All Strange Away,* 169.

52. Darren Gribben, "Samuel Beckett: Number 465. Censorship of the Self and Imagination in Beckett's Work after World War II," in Marius Buning et al., eds., "Three Dialogues Revisited/Les Trois dialogues revisités," *Samuel Beckett Today/Aujourd'hui* 13 (Amsterdam: Rodopi, 2003), 221. Gribben, furthermore, astutely argues that *Ping* and *Lessness* "are influenced by Beckett's experiences of World War II; that they literally encrypt the real story, hide it from view" (216), and that *All Strange Away,* in which war is mentioned, is about imprisonment and torture. Gribben's analysis of the prose fiction opens up new ways to read these texts.

53. Beckett, *All Strange Away,* 169. Emphasis added.

54. Ibid.

55. I have omitted page numbers when the repetition is so frequent that to list all or nearly all the pages in these short texts would seem redundant.

56. Samuel Beckett, *Ping,* in *The Complete Short Prose,* 193.

57. Ibid.

58. Beckett, *All Strange Away,* 171.

59. Ibid., 172–73.
60. John Pilling and James Knowlson, *Frescoes of the Skull: The Later Prose and Drama of Samuel Beckett* (London: John Calder, 1979), 139; Peter (P.J.) Murphy, *Reconstructing Beckett: Language for Being in Samuel Beckett's Fiction* (Toronto: University of Toronto Press, 1990), 86; Graham Fraser, "The Pornographic Imagination in *All Strange Away*," *Modern Fiction Studies* 41.3–4 (1995): 515–30.
61. Samuel Beckett, *Lessness,* in *The Complete Short Prose,* 198.
62. Samuel Beckett, "The End," in *Stories and Texts for Nothing* (New York: Grove Press, 1967), 61.
63. Perloff, "In Love with Hiding," 93.
64. For a full account of the living conditions of the inhabitants of Saint-Lô, see Gaffney, *Healing Amid the Ruins.*
65. Beckett, *Lessness,* 200.
66. Ibid., 199.
67. Ibid., 200.
68. Ibid., 199 and 200.
69. Beckett, *Disjecta,* 145.
70. Beckett, *Lessness,* 199.

CHAPTER 9

Writing Relics: Mapping the Composition History of Beckett's *Endgame*

Dirk Van Hulle

Retrospection inevitably tends to present past events in terms of later ones. Manuscript research and genetic criticism are not immune to this powerful impulse. Presented with a complex set of writing relics, genetic critics are supposed to establish a chronology and reconstruct the writing history—according to the motto "make what sense you can of that."[1] During this reconstruction it may be tempting to project dramatic structures into the writing process in order to be able to present the published text as the dénouement or the inevitable outcome of a linear process, narratable by means of flashbacks and foreshadowing. In this respect, it may be useful to take the advice of critics who have studied the interface between historiography and narratology, such as Michael André Bernstein, whose book, *Foregone Conclusions: Against Apocalyptic History,* and its concepts of fore-, back-, and side-shadowing, will serve as guidelines in this investigation into the interplay of history, memory and the archive in Beckett studies.[2] Because of its long composition history, *Fin de partie / Endgame* is a suitable case study to examine Beckett's "travail"—as he preferred to call his writing, regarding his work as a verb rather than a noun.

Two categorizations—one by a poet and one by a critic—may serve as a useful starting point. In "Die Entstehung eines Gedichts," Hans Magnus Enzensberger distinguishes between two ways of approaching a poem's composition history: from the inside and from the outside, noting that only the author can study the composition process "from the inside."[3] The other approach can only start from what Enzensberger calls "unequivocal" (*eindeutig*) yet "poor" material—"poor" because there are no memories attached to it. This is the material genetic criticism has to work with: notes, marginalia, jottings, drafts, typescripts, proofs, self-translations, theatrical notebooks. Its research focus is the dynamics of the writing process insofar as it can be reconstructed on the basis of its material traces, which are not always that "unequivocal."[4] As a creative artist, Enzensberger naturally tends to approach the genesis of his work from the first perspective.

The second point of view is reflected in Pierre-Marc de Biasi's categorization. Because of the material starting point of genetic criticism, Pierre-Marc de Biasi conceived a basic model of the average writing process as a typology of genetic documentation.[5] His model hinges on the so-called "pass for press" moment (*bon à tirer*, i.e. the moment the author decides that all is set for printing), and is a general typology, applicable to various collections of textual material. De Biasi is well aware of the artificial nature of his undertaking, since every writing process is characterized by singularities that deviate from the general typology, but in general he distinguishes three different phases of the writing process: pre-composition (exploration, documentation, reading, note taking, conceptualization); composition (textualization, drafts, typescripts, revisions); and post-composition (publication history, performance history, self-translations). The moment the author decides his text is ready to be printed/published (*bon à tirer*) plays a central role in de Biasi's scheme, but in Beckett's case, of course, that moment is complicated by his practices of self-translating and directing his own plays. This emphasizes that the publication of each of Beckett's works is an endgame—nearly, but never quite, finished— and that each version is inextricably bound up with its textual history. To examine this endgame-like nature of Beckett's bilingual works, the manuscripts witnessing to the composition history of *Fin de partie / Endgame* will serve as a suitable corpus.

Foreshadowing

The reconstruction of a text's genesis inevitably implies interpretation. Although the empirical basis is the starting point of genetic research, the *avant-texte* is more than the sum total of all extant material documents. Samuel Beckett donated many of his manuscripts to university archives, but these are only the relics of a process, driven by impulses that are no longer present. The task of genetic criticism is not just to quietly study the manuscripts, but also to examine the dynamics of the writing process and to visualize this composition history, for instance by means of electronic tools. In this way the manuscripts do not simply remain the relics of writing, but become the fascinating witnesses of the way Beckett often conceived of his writings as relics (*Residua, Disjecta,...*) with a textual history.

The problem with the reconstruction of such a textual history is the danger of foreshadowing. Since most readers of Beckett have come to appreciate his texts by means of their published versions, it is only natural that these texts' composition histories tend to be reconstructed with a view to understanding their "definitive" state. This teleology is to a certain extent justified, because any writing process implies some kind of project. At the same time, the "telos" or final end of the project is seldom crystal-clear from the very start. Moreover, authors such as Samuel Beckett, who are extremely conscious of their own creative processes, often explicitly question the possibility of ever reaching this end. Especially in Beckett's later works, the textual difficulty of closure coincides with the existential impossibility of "effing" the "ineffable departure." Thus, for instance, his penultimate text can only end with a wish to end ("Oh all to end") and his last text is deliberately left unfinished. This intentional incompletion has a retroactive effect on Beckett's whole *oeuvre,* which he chose to end in the middle of a sentence with the words "comment dire."

With hindsight it may seem as if *Endgame*'s first spoken words already foreshadow this open ending of his whole *oeuvre:* "Finished, it's finished, nearly finished, it must be nearly finished."[6] But this is a form of foreshadowing. In *Foregone Conclusions,* Bernstein notes that in its most extreme form "foreshadowing implies a closed universe in which all choices have already been made," whether it happens at a theological, historical, or psychological level: "Christian apologetics, Marxist teleology, and psychological determinism are striking instances of how powerful our impulse toward foreshadowing can be."[7] This impulse is especially noticeable in autobiographies and personal accounts of writing processes, that is Enzensberger's first category. To illustrate how wary Beckett is of this impulse, two accounts by other authors may serve as a contrastive background. For instance, in the 1831 introduction to her novel *Frankenstein, or: the Modern Prometheus,* Mary Shelley retrospectively reconstructs the genesis of her "hideous progeny." When, during the "wet, ungenial summer" of 1816, Lord Byron invites each of his guests to write a ghost story, Mary initially can't think of anything: "I felt that blank incapability of invention which is the greatest misery of authorship, when dull Nothing replies to our anxious invocations."[8] Later on, she realizes that in this way she could have been waiting for inspiration forever, because "Invention, it must be humbly admitted, does not consist in creating out of void, but out of chaos." She accidentally comes across the raw material for her story when she overhears a conversation between Byron and Percy Shelley on "the nature of the principle of life."[9] After this conversation, Mary Shelley—according to her account—can't sleep. She continues the narrative of the creative process in a tone that comes close to that of a horror story: "I saw the hideous phantasm of a man stretched out, and then, on the working of some powerful engine, show signs of life, and stir with an uneasy, half-vital motion [...]. The idea so possessed my mind, that a thrill of fear ran through me [...]." And then she experiences—again, according to the retrospective account—a kind of Eureka moment: "I have found it!"[10]

A similar *eureka* moment characterizes Charles Darwin's retrospective account of the writing process in his *Autobiography,* which is a typical example of foreshadowing. He mentions his reading of Malthus' *Essay on the Principles of Population* in October 1838 and presents it as a kind of intellectual lightning that suddenly strikes him: "being well prepared to appreciate the struggle for existence [...] it at once struck me that under these circumstances favourable variations would tend to be preserved, and unfavourable ones to be destroyed. The result of this would be the formation of new species. Here then I had at last got a theory by which to work."[11] With hindsight, the writing process seems to have been a straight line of textual descent, but from a chronological point of view it could have developed in numerous other directions. Howard Gruber points out that this insight did not come as a *eureka* moment. "As it happens, the crucial passage does not even contain a single exclamation point, although in other transported moments he [Darwin] used quite a few, sometimes in triplets."[12] Darwin's account is, of course, an understandable phenomenon. In genetic criticism this teleological viewpoint has given rise to hot debates. According to Almuth Grésillon, "the teleological view perverts the interpretation, makes it blind to accidental events, to losses, to the state of suspension, to open alternatives, in other words to all forms of writing that deviate from the straight course."[13]

Darwin demonstrated that man is not the centre of biological phenomena. The human race is not the "telos" of creation, but just one of the countless twigs on what he called the "tree of life." This move away from anthropocentrism is already present in an early notebook on transmutation (the so-called B-notebook), started in the summer of 1837, more than twenty years before the first edition of *The Origin of Species.* But at the end of his life, looking back on the development of his own theory, it may have been almost impossible for him not to see the development of his own theory as a straight path toward the telos. And foreshadowing not only occurs in the approach "from within"; the impulse toward foreshadowing appears to be just as powerful in Enzensberger's second category. In retrospect, Darwin's son Francis (in his 1909 introduction to his father's 1842 and 1844 essays) was astonished "that Malthus should have been needed to give him [Charles Darwin] the clue," for a "forecast of the importance of the survival of the fittest" was already present in the 1837 notebook.[14] But obviously Charles Darwin had to formulate his theory first before Francis could notice the "forecast." The notion of "forecast" is symptomatic of the foreshadowing that characterizes not only his, but many other accounts as well. Moreover it contains a subdued and possibly unintentional reproach, which turns this "foreshadowing" into a mild form of "backshadowing."

Backshadowing

The term "backshadowing" is coined by Bernstein to denote "the most pervasive, but also the most pernicious, variant of foreshadowing."[15] "Backshadowing is a kind of retroactive foreshadowing in which the shared knowledge of the outcome of a series of events by narrator and listener is used to judge the participants in those events *as though they too should have known what was to come.*"[16] In the above example, Francis Darwin's account of the writing process of *The Origin of Species* involuntarily sounds as if he cannot believe his father should have had recourse to Malthus to see the obvious. Still, he does have a point in that he refuses to believe in the retrospective tendency to present a mental development as a sudden revelation. And it is clear that Francis Darwin did not intend to present his father as a man without genius. If anything, the opposite is true. With reference to the genesis of *The Origin of Species,* the first supportive readers' judgmental act of backshadowing is usually not directed at Charles Darwin, but at themselves. In a letter to Darwin, H.C. Watson writes: "Now these novel views are brought fairly before the scientific public, it seems truly remarkable how so many of them could have failed to see their right road sooner."[17] One of the most famous instances of "self-backshadowing" was Thomas Henry Huxley's reaction: "How extremely stupid not to have thought of that!"[18]

In genetic Beckett studies, and the study of *Fin de partie / Endgame* in particular, an interesting case of foreshadowing, including a mild form of backshadowing, is Giuseppina Restivo's essay "The Genesis of Beckett's *Endgame.*" This essay, based on important philological spadework and manuscript research, traces the "core" and the "basis" of the play to a few drafts in the so-called "Sam Francis" notebook.[19] It contains two dialogues between "A" and "B"[20] and one dialogue between "X" and his factotum "F."[21] In the first few (cancelled) lines of the first fragment, A and B respectively persuade themselves that they are not alone. In the rest of the fragment, A and B seem to be preoccupied with the presence of a third (B has the feeling there is someone in the neighborhood.[22] But their mutual presence is equally uncertain. In both this and the next fragment the characters wonder whether the "third" might be Christ. In the second A-B dialogue (dated "15 septembre 1950 / retour d'Irlande"), B comes to the conclusion that there is nobody. But that realization is of no help as long as he has the feeling that someone's there.[23]

To refer to this pair of fragments Giuseppina Restivo uses the abbreviation "*AV.1*" because they "belonged to the same *Avant Fin de partie* phase."[24] "*Avant Fin de partie*" was the label Beckett retroactively wrote on the typescript of the X-F dialogue.[25] This label is one of the few hints of foreshadowing in Beckett's *oeuvre*, though it is not the only manuscript he has given the "avant" label. For instance, the theatrical fragment "Good Heavens" was labeled "Before *Come and Go*" when Beckett prepared it "For Reading University Library."[26] Although the fragment shows several structural correspondences with *Come and Go*, its text differs significantly. The "*Avant Fin de partie*" label is a similar case of mild foreshadowing. Only after the completion of *Fin de partie* could Beckett designate the typescript as part of the play's composition history. Restivo's abbreviation reflects her thesis that in this (pre)history the A-B dialogues and the X-F dialogue in the Sam Francis notebook together form the "two different starts" of *Endgame*'s "line of development."[27]

The catalogue *Beckett at Reading* is more cautious. Under the heading "Early versions of *Fin de partie*" it specifies: "While the fragments in the MS 2926 notebook cannot be described as an early draft of the play, they do reveal links with later versions of *Fin de partie*."[28] In spite of the many differences (in tone, length, structure…) "these fragments do contain, in a very elementary form, many of the preoccupations of the final text."[29] The confusion about the status of these fragments is reflected in the different terms that are employed to describe them. According to the catalogue, they are "versions" but not "drafts" of *Fin de partie*; according to Restivo they are the "first start";[30] according to Ruby Cohn they are a "step" toward *Fin de partie* and "the origin of the stemma"—even though Cohn believes that "Restivo exaggerates the 'ending' quality of those fragments."[31]

A Beckett Canon is, among many other things, an admirable combination of a bird's-eye view on Beckett's *oeuvre* as a whole with Beckett's own perspective on his texts at the time he was writing them: "Without a thesis to uphold, I will wend my way chronologically through Beckett's works, gliding as he did from one genre to another, from one theme to another, from one wordscape to another."[32] In this way, Beckett's writings do not appear as a static *oeuvre*, but as a work in motion. Thanks to the chronological structure of this attempt to map Beckett's works, including the manuscripts, Cohn's canon can be regarded as a rare attempt to apply the idea of a "genetic dossier"[33] to the entirety of an author's works. Building on the pioneering work by S.E. Gontarski, Ruby Cohn suggests an alternative stemma for *Fin de partie / Endgame,* which differs from Restivo's "line of development"[34] and deviates from the descriptions in the catalogue in that not all fragments that are catalogued as "Early versions of *Fin de Partie*" are included in Cohn's stemma. For instance, the dialogue between "Ernest et Alice" is described in the Reading catalogue as a dialogue that foreshadows *Endgame,* a dialogue "between a Hamm-like Ernest, confined not to a wheelchair but to a cross, and his wife Alice, who ministers to his needs," whereas Ruby Cohn regards it as "an odd fragment, unrelated to *Fin de partie.*"[35]

These disagreements among different researchers illustrate the problematic issue of foreshadowing in genetic criticism. What the Catalogue of the Beckett Manuscript Collection refers to as "abandoned dramatic material"[36] is part of the *avant-texte* of *Fin de partie,* according to Restivo, who corroborates her thesis with several examples of fragments that "foreshadow" elements in the "final text" of the play: "The expectation that somebody will come [...] may very likely *foreshadow* the final appearance of the boy in *Endgame*"; "A's fear that B will fail to answer prepares for Clov's threat of departure, of leaving Hamm alone and unanswered"; "Such obvious religious nostalgia and the characters' psychological needs [...] stand at the basis of the final text."[37]

While Restivo presents as many examples as possible to corroborate her argument, the enumeration of correspondences has the involuntary side effect of backshadowing: the longer the list becomes, the more it may give the impression of a silent reproach to Beckett: why did it take him so long to write *Fin de partie* if he already had most of the play's ingredients in 1950? This side effect is clearly unintentional, but it is nonetheless one of the dangers of a teleological approach.

Sideshadowing

As opposed to back- and foreshadowing, "sideshadowing champions the incommensurability of the concrete moment and refuses the tyranny of all synthetic master-schemes; it rejects the conviction that a particular code, law, or pattern exists, waiting to be uncovered beneath the heterogeneity of human existence."[38]

The "avant-texte" of *Fin de partie* already illustrated that different researchers include different holograph manuscripts under this label. The "Kilcool" fragments are another example.[39] As S.E. Gontarski and Rosemary Pountney have shown, the correspondences between the "Kilcool" fragments and *Not I / Pas moi* are striking, but it remains difficult to determine whether one could go as far as to regard these fragments as "versions" of *Not I*.[40] The "Kilcool" fragments are more than just relics of writing; they also exemplify the act of writing relics, that is texts that were meant to be relics from the start, in the etymological sense of the Latin verb *relinquere*, to leave behind—which accords with the content of the first "Kilcool" fragment. The girl (who moves to Kilcool to go and live with her aunt) is left behind. Her parents have died. Pray as she may to Time the Father[41] she cannot join her parents. In case she might consider committing suicide, her aunt tries to reassure her.[42] Not unlike Goethe's Faust after his attempted suicide, when he has to conclude "Die Erde hat mich wieder," the girl can only come to a similar conclusion, but with a Beckettian twist—as in his addenda to *Watt:* "Die Merde hat mich wieder."[43] The additional M indicates the link between the idea of being left behind and leaving something behind in the form of writing relics. Writing as a form of excretion is a persistent theme throughout Beckett's work. For instance in *Not I*, "the wild stream of words is expressly linked by Mouth with excremental discharge, 'nearest lavatory... start pouring it out... steady stream... mad stuff... half the vowels wrong,'" as James Knowlson and John Pilling note in *Frescoes of the Skull*.[44] The "uncontrollable urge to pour out words" can be linked to "the obligation to express"—as Beckett formulated this creative urge in the *Three Dialogues with Georges Duthuit*—while the result of this discharge is called "verbal diarrhoea"[45] or "logorrhea."[46]

With the (dis)advantage of hindsight it is difficult to imagine Beckett's writing situation in 1950; how his intentions could change; and how variation on a theme of one work could gradually lead to the creation of a new work. In this context, it is important that both Restivo's essay and the catalogue *Beckett at Reading* also mention the resemblances between the "A et B" fragments and *En attendant Godot* (apart from the correspondences with *Fin de partie*).[47]

The "A et B" fragments are an intermediate stage between *En attendant Godot* and *Fin de partie*. Since the former was already "finished," it makes sense to present them as precursors of the latter, but at the same time they belong to "neither"[48]—which is such a Beckettian situation that it is worthwhile thinking of alternative methods to present such manuscripts. On the one hand *En attendant Godot* was published, but only "finished" in the sense of an endgame, as in Clov's opening line "Finished, it's finished, nearly finished, it must be nearly finished."[49] On the other hand the "A et B" fragments show resemblances not only with *Fin de partie,* but with several other works as well. For instance, B's remark that the feeling of being observed is disgusting[50] corresponds with the Berkeleyan subtext of *Film*. In the second fragment, the stage direction "paroles inintelligibles"[51] preceding the opening of the play shows some resemblance with "*MOUTH's voice unintelligible behind curtain*" before the opening of *Not I*.

Apart from these correspondences between a manuscript and published works, there are also correspondences between manuscripts from quite different periods; thus, for instance, the "organ" in the 1950 fragment (in which one of the characters urges the other to make a longer statement in order to get accustomed to his organ)[52] is still a preoccupation in one of the abandoned drafts of *Stirrings Still* (dating from 1983–1984), where "they" come and read to the old protagonist, who can only discern between them thanks to the difference in timbre and accent of their respective organs.[53]

Sideshadowing implies an increased awareness of the impulse toward foreshadowing as a way of creating meaning. Translated to the "*avant-texte*" of literary texts, this implies the realization that the notion of "*avant-texte*" itself is already a problematic concept because it derives its meaning from its opposition to the notion of the "text." According to Jean Bellemin-Noël, who coined the term in 1972, it denotes all the material evidence of what precedes the moment when a work is finally "treated as a text."[54] Pierre-Marc de Biasi, however, defines the *avant-texte* as the result of the critical analysis and organization of the extant documents relating to the writing process one wishes to examine.[55]

Still, this is not enough to avoid the impulse toward foreshadowing. Ruby Cohn includes the early fragments in her stemma, yet in an endnote she adds: "I find little of *Fin de partie* in these fragments, but I assign them to the stemma as a matter of convenient collation with B[eckett] at R[eading]."[56] An alternative would have been to treat them as "odd fragments" like "Ernest et Alice," but—Cohn adds in the same endnote—"Restivo (1994) is most insistent on the *seed* of *Fin de partie* in the 1950 scene."[57]

The "seed" is an interesting metaphor, suggesting an organic notion of the work's genesis, which seems applicable to Restivo's view, as she speaks of the "long creative process during which *Endgame* was *born*."[58] The problem with this organic metaphor is the implicit assumption that the "seed" or "core" of a work of literature already contains all its ingredients *in nuce* and is predestined to "grow" into the shape it eventually took. This problem is closely linked to nomenclature and classification. In Ruby Cohn's *A Beckett Canon* manuscripts such as "Ernest et Alice," "Kilcool," or "Last Soliloquy" are presented as separate (unpublished) works with a title of their own, while for instance "Good Heavens" is not. In biological classification models, such as Darwin's "tree of life," they would be presented as dead ends. In such a model the "A et B" fragments could be interpreted as the origin of a new branch or textual species rather than "the origin of the stemma."[59]

Conclusion

Perhaps the term "stemma" is not so suitable to denote a linear reconstruction or "line of development."[60] If a stemmatic model is suitable at all to map and visualize the composition history of Samuel Beckett's works, it might be worth considering to apply it to Beckett's *oeuvre* as a whole, including both manuscripts and published works, all of Beckett's writing relics. This way Beckett's work (in the sense of "travail" rather than "*oeuvre*") would appear as a groping process—not in the sense that Beckett did not know what he was doing, but in the sense that he did not always have a clear idea yet of what a new text was going to look like.

In the 1980s and 1990s, a fashionable alternative to the arboreal, stemmatic model was the rhizome, especially since its introduction by Gilles Deleuze and Félix Guattari in *Mille Plateaux*. Perhaps the rhizome metaphor is applicable to the "underground" nature of sideshadowing references such as the "organ" correspondence between the 1950 and 1983–1984 manuscripts which eventually did not make it into print. Again, this is just a metaphor, but nonetheless one of the "metaphors we live by" (as Lakoff and Johnson called them),[61] reflecting the ways in which we view the world, and at the same time influencing our world views. In this sense, the metaphor of the rhizome was presented by Deleuze and Guattari as an alternative to a traditional, hierarchical, "arborescent" way of thinking.[62] During the early 1990s, this central metaphor in the hypertext euphoria was linked back to the essay "As We May Think" by Vannevar Bush and his Memex idea, regarded as one of the main precursors of the internet.[63]

Now that the hypertext euphoria is over, the arboreal and rhizomatic metaphors no longer obligatorily need to be presented as reflecting two diametrically opposed ways of thinking. This is where the notion of sideshadowing may be useful: "It is between the antithetical but twin reductionisms of teleological determinism and radical undecidability, between, to give as examples two of their recent local names, Marxism and Deconstruction, that the prosaics of sideshadowing positions itself," Bernstein argues.[64] One of his best examples to illustrate these "prosaics" is *A la recherche du temps perdu*. When Samuel Beckett wrote his essay on Marcel Proust at the start of his career, he was already well aware of what Bernstein calls "the tense equilibrium between epistemological uncertainty and analytic assertiveness," which according to him is "paradigmatic for the more general question of how we try to reconcile the competing attractions of foreshadowing and sideshadowing in our own lives, both as readers of texts and as interpreters of our own histories."[65] In Beckett's case, this awareness is reflected in the numerous intratextual, sideshadowing references between his own works.

To some extent the need to classify and pigeonhole manuscripts is imposed upon researchers by the medium of printed books. Scholarly editions on paper cannot but cut off the writing process of a particular work from the entirety of Beckett's writings. My suggestion is that it would be in the spirit of the "continuing incompletion,"[66] characterizing many of Beckett's works, to treat his whole *oeuvre* as one long textual genesis, in which the *avant-texte* of *Fin de partie* is only one module. A digital manuscript project[67] can be a flexible tool to map the composition history of Beckett's texts, including facsimiles and transcriptions of the documents, many of which were donated by the author to university libraries. Scanning these documents is not only useful for preservation purposes; it also creates an alternative to foreshadowing, especially with regard to manuscripts that are hard to classify, such as the "A et B" and "Kilcool" fragments.

The advantage of sideshadowing in a digital environment is that it enables researchers to link pieces of information to any other kind of data. In the meantime this has become common practice: whoever uses the internet immediately adapts to its hypertextual structure: any piece of information in the gigantic network of hyperlinks momentarily becomes the center of the web when it appears on one's screen. Applied to Beckett's manuscripts this implies that, in a digital manuscript archive or a genetic edition, any document can momentarily become the center of Beckett's *oeuvre*. It does not need to be presented in a subordinate position to the published versions of *Fin de partie*. It can be presented the way it was in front of Beckett's eyes while he was putting pen to paper. This way of presenting enables readers to follow Cohn's example, "gliding as he [Beckett] did [...] from one wordscape to another,"[68] and to put themselves into Beckett's place at the moment this manuscript could still potentially develop into numerous directions.[69] On a macro-level Beckett's writings as whole thus appear as both a "work in progress" and a "work in regress"—as he called it himself. And on a micro-level the writing relics can come out of the (fore)shadows in a digital manuscript project that makes it possible to present them as an archival equivalent of what Beckett described as *"neither"* and to "gently" shed some light on their in-between position—"unfading on that unheeded neither."[70]

Notes

1. Samuel Beckett in a letter to Pamela Mitchell, dated 6 August 1954. Quoted in James Knowlson, *Damned to Fame: The Life of Samuel Beckett* (London: Bloomsbury, 1996), 403.
2. Michael André Bernstein, *Foregone Conclusions: Against Apocalyptic History* (Berkeley: University of California Press, 1994).

3. Hans Magnues Enzensberger, "Die Entstehung eines Gedichts," in *Gedichte. Die Entstehung eines Gedichts* (Frankfurt am Main: Suhrkamp, 1962), 62.

4. Ibid., 62.

5. Pierre-Marc de Biasi, "What is a Literary Draft? Toward a Functional Typology of Genetic Documentation," *Yale French Studies* 89 (1996): 26–58.

6. Samuel Beckett, *Endgame,* in *The Complete Dramatic Works* (London: Faber and Faber, 1990), 93.

7. Bernstein, *Foregone Conclusions,* 2.

8. Mary Shelley, *Frankenstein, or the Modern Prometheus* (London: Penguin Classics, 1992), 8.

9. Ibid., 8.

10. Ibid., 9.

11. Charles Darwin, *Autobiography*, in Francis Darwin, *The Life of Charles Darwin* (London: Studio Editions, 1995), 40.

12. Howard E. Gruber, *Darwin on Man: A Psychological Study of Scientific Creativity* (London: Wildwood House, 1974), 170.

13. "Le regard téléologique pervertit l'interprétation, la rend aveugle à l'accident, à la perte, à l'état de suspension, à l'alternative ouverte, bref à toutes ces formes d'écriture qui s'écartent de la ligne droite." Almuth Grésillon, *Eléments de critique génétique: Lire les manuscrits modernes* (Paris: Presses Universitaires de France, 1994), 138.

14. Francis Darwin, ed. *The Foundations of the Origin of Species: Two Essays Written in 1842 and 1844* (Cambridge: Cambridge University Press, 1909), xv–xvi.

15. Bernstein, *Foregone Conclusions,* 16.

16. Ibid., 16. Bernstein's italics.

17. Quoted in Francis Darwin, *The Life of Charles Darwin* (London: Studio Editions, 1995), 168.

18. Quoted in Tim Lewens, *Darwin* (New York: Routledge, 2007), 46.

19. Giuseppina Restivo, "The Genesis of Beckett's *Endgame* Traced in a 1950 Holograph," in Marius Buning et al., eds. "Intertexts in Beckett's Work et/ ou Intertextes de l'oeuvre de Beckett," *Samuel Beckett Today/Aujourd'hui* 3 (Amsterdam: Rodopi, 1994), 85; 89.

20. UoR MS 2926 f. 6r–10r; 11r–20r. © Samuel Beckett Estate.

21. UoR MS 2926 f. 23v–48r. © Samuel Beckett Estate.

22. UoR MS 2926 f. 7r. © Samuel Beckett Estate.

23. UoR MS 2926 f 12r. © Samuel Beckett Estate.

24. Restivo, "Genesis of Beckett's *Endgame,*" 85.

25. UoR MS 1227/7/16/7. © Samuel Beckett Estate.

26. UoR MS 1227/7/16/4. © Samuel Beckett Estate.

27. Restivo, "Genesis of Beckett's *Endgame,*" 85; 93.

28. Mary Bryden, Julian Garforth, and Peter Mills, *Beckett at Reading: Catalogue of the Beckett Manuscript Collection at the University of Reading* (Reading: Whiteknights Press and the Beckett International Foundation, 1998), 33–34.

29. Ibid., 34.

30. Restivo, "Genesis of Beckett's *Endgame*," 94.

31. Ruby Cohn, *A Beckett Canon* (Ann Arbor: University of Michigan Press, 2001), 220–221.

32. Ibid., 1–2.

33. Almuth Grésillon suggests this term as an alternative to the more common term *"avant-texte"* and defines it as "un ensemble constitué par les documents écrits que l'on peut attribuer dans l'après-coup à un projet d'écriture determine dont il importe peu qu'il ait abouti ou non à un text publié." *Eléments de critique génétique,* 109.

34. For instance, unlike Ruby Cohn, Giuseppina Restivo does not regard the four-leaf dialogue between X and F in Trinity College, Dublin (TCD MS 4662) as a separate version preceding the X-F dialogue in the Sam Francis notebook, but as the second part of one single holograph version (UoR MS 2926, f. 23v–48r + TCD MS 4662, f. 1r–4r) preceding the *"Avant Fin de partie"* typescript. "Genesis of Beckett's *Endgame*," 85.

35. Cohn, *A Beckett Canon,* 220.

36. Bryden et al., *Beckett at Reading,* 195.

37. Restivo, "Genesis of Beckett's *Endgame*," 87; 88. Emphasis added.

38. Bernstein, *Foregone Conclusions,* 4.

39. Trinity College, Dublin, MS 4664, ff. 10r–19r. © Samuel Beckett Estate.

40. S.E. Gontarski, *The Intent of Undoing in Samuel Beckett's Dramatic Texts* (Bloomington: Indiana University Press, 1985), 131–49; Rosemary Pountney, *Theatre of Shadows: Samuel Beckett's Drama 1956–76* (New Jersey: Colin Smythe/Barnes and Noble Books, 1988), 92–102.

41. Trinity College, Dublin, MS 4664, f. 10r. © Samuel Beckett Estate.

42. Trinity College, Dublin, MS 4664, f. 11r. © Samuel Beckett Estate.

43. Samuel Beckett, *Watt* (London: Calder, 1981), 251.

44. James Knowlson and John Pilling, *Frescoes of the Skull: The Later Prose and Drama of Samuel Beckett* (London: Calder, 1979), 200.

45. Ibid., 200.

46. Gontarski, *The Intent of Undoing,* 133.

47. Restivo, "Genesis of Beckett's *Endgame*," 87; Bryden et al., *Beckett at Reading,* 34.

48. Samuel Beckett, *The Complete Short Prose, 1929–1989,* ed. with notes and an introduction by S.E. Gontarski (New York: Grove Press, 1995), 285.

49. Beckett, *Endgame,* 93.

50. UoR MS 2926, 7r. © Samuel Beckett Estate.

51. UoR MS 2926, 11r. © Samuel Beckett Estate.

52. UoR MS 2926, 9r. © Samuel Beckett Estate.

53. UoR MS 2935/1/2. © Samuel Beckett Estate.

54. Jean Bellemin-Noël, *Le texte et l'avant-texte: Les brouillons d'un poème de Milosz* (Paris: Librairie Larousse, 1972), 15.

55. Biasi, "What is a Literary Draft?," 30–31.

56. Cohn, *A Beckett Canon,* 401n3.

57. Ibid. Emphasis added.
58. Restivo, "Genesis of Beckett's *Endgame,*" 90. Emphasis added. Ruby Cohn employs a similar metaphor with reference to the "Kilcool" fragments: "Even aborted, 'Kilcool' proved pregnant for Beckett." *A Beckett Canon,* 280.
59. Ibid., 220.
60. The notion of the stemma (in the sense of a family tree of texts), as it was developed by Karl Lachmann and others in the first half of the nineteenth century, was usually applied to the copying of medieval manuscripts or the distribution of texts in different editions.
61. George Lakoff and Mark Johnson, *Metaphors We Live By* (Chicago: University of Chicago Press, 1980).
62. Gilles Deleuze and Félix Guattari, *Mille plateaux: capitalisme et schizophrénie* (Paris: Les éditions de minuit, 1980), 20.
63. Vannevar Bush, "As We May Think," *The Atlantic Monthly*, July 1945, http://www.theatlantic.com/unbound/flashbks/computer/bushf.htm>.
64. Bernstein, *Foregone Conclusions,* 7.
65. Ibid., 117.
66. H. Porter Abbott, *Beckett Writing Beckett: The Author in the Autograph* (Ithaca and London: Cornell University Press, 1996), 20.
67. For a description of the Beckett Digital Manuscript Project, see Dirk Van Hulle and Mark Nixon, "Holo and unholo," in Dirk Van Hulle and Mark Nixon eds., "All Sturm and no Drang," *Samuel Beckett Today/Aujourd'hui* 18 (Amsterdam: Rodopi, 2007), 313–22.
68. Cohn, *A Beckett Canon,* 1–2.
69. In this context it is interesting to note that Ruby Cohn emphasizes the potential of the "Kilcool" fragments by describing them as "four outlines for a possible play." *A Beckett Canon,* 280.
70. Beckett, *Complete Short Prose,* 258.

CHAPTER 10

"Agnostic Quietism" and Samuel Beckett's Early Development

Matthew Feldman

"All I ever got from the *Imitation* [*of Christ*, by Thomas á Kempis] went to confirm and reinforce my own way of living," wrote a confessional Samuel Beckett to his closest friend, Thomas MacGreevy, on 10 March 1935, "a way of living that tried to be a solution and failed." Despite finding "quantities of phrases [...] that seemed to be made for me and which I have never forgotten," he continued:

> What is one to make of "seldom we come home without hurting of conscience" and "the glad going out and sorrowful coming home" and "be ye sorry in your chambers" but quietism of the sparrow alone upon the housetop and the solitary bird under the eaves? An abject self-referring quietism indeed, beside the alert quiet of one who always had Jesus for his darling, but the only kind that I, who seem never to have had the least faculty or disposition for the supernatural, could elicit from the text, and then only by means of a substitution of terms very different from the one you propose. I mean that I replaced the plenitude that he calls "God" not by "goodness," but by a pleroma only to be sought among my own feathers or entrails, a principle of self the possession of which was to provide a rationale and the communion with which a sense of grace [...] I know that now I would be no more capable of approaching its hypostasies and analogies "meekly, simply and truly," than I was when I first twisted them into a programme of self-sufficiency.[1]

Beyond the aesthetic and self-referential qualities of this insightful passage, the basis for starting out with Beckett's letter centers upon the fundamental idea propounded both there and here: "agnostic quietism." If a lowest common denominator—agnosticism about how words, experience or reality, in addition to an all-too-familiar mental anguish—may be said to be perceptible in Beckett's work, how does one go about characterizing it? Surely by *The Trilogy* and *Endgame*, agnosticism confronts Beckett's creatures in its baldest form: Does God exist and therefore bear responsibility for suffering? Does the question itself matter, or offer any succor? Can truth be apprehended, even briefly, in the first place? Indeed, how do "I"—whatever that means—form the words to express to others and myself such sentiments? And to what end? This chapter will argue that quietism provides an ethical and, for Beckett, aesthetic approach to suffering and failure *as a spiritual purgation for living;* and that this "agnostic quietism," in turn, provides an appropriate heuristic torch to shine into his wan writings.

Aside from Mark Nixon's insightful account of Beckett's interest in the "quietistic and pessimistic tradition" in "Scraps of German" and James Knowlson's biographical views on Beckett's "quietistic impulse," the only other Anglophone text to consider quietism in any depth is by Chris Ackerley.[2] Ackerley rightly defines quietism as a "doctrine of extreme asceticism and contemplative devotion teaching that the chief duty of man is the contemplation of God, or Christ, to become independent of outward circumstances and sensual distraction."[3] Moreover, it is hard to contest Ackerley's dating and conclusions, and we shall take his understanding of quietism—a seventeenth century quasi-Gnostic doctrine of Christianity ultimately condemned as heretical—as definitional (minus the Christian piety for Beckett, of course). But as Beckett's writings make clear, his own "agnostic quietism" is not passively skeptical. For he says elsewhere in the same letter to MacGreevy that the ideals of "humility, utility, self-effacement," when deprived of a Christian teleology, become a way of suffering life without hope of compensation—everlasting or earthly.[4] Thomas à Kempis' *The Imitation of Christ* offers a fitting segue, then, not only because of Ackerley, Knowlson and Nixon's important insights, nor simply because Beckett's 1931 reading of this text contributed immensely to his letter to MacGreevy on quietism, but also because quietism forms part of a larger current in his formative readings and intellectual development. If we recall Beckett's description of a "mind that has raised itself to the grace of humility" in his 1934 tribute to MacGreevy,[5] or his idea that (good) poetry approaches prayer—in sentiments both for MacGreevy and to himself[6]—it is possible to begin to characterize the kind of "agnostic quietism" suffusing Beckett's writing during 1930s and beyond. When constructing a reading of Beckett's art in terms of the suffering, uncertainty, failure, and stoicism that are so often held to be thematically present in his texts, a prior accumulation of his sources on such matters will reveal the ways in which Beckett used erudition to explore unknowing.

In the first place, it is important to bear in mind that the roots of Beckett's "sad" outlook predates his reading of the *Imitation of Christ* in late summer 1931, and extend well beyond his self-analysis to MacGreevy in 1935. For example, Ackerley carefully traces Kempis beyond the thirty-plus allusions in *Dream of Fair to Middling Women* and persistent references in *More Pricks than Kicks,* to residua in *Watt, The Trilogy* and *How It Is,* despite being less concerned with the development of Beckett's own outlook than with charting the references of Kempis in Beckett's texts. Yet Ackerley only cursorily mentions Arthur Schopenhauer as a quietistic companion of Kempis, yet he too may be seen to underpin Beckett's voluminous readings into this subject. Thus taking the influence of Kempis as given, a survey of the other texts helping to hatch this "solitary bird under the eaves" is in order. This overview will, of necessity, be both genealogical and intuitive—built on what Johann Goethe called "elective affinities"—as only in this way can we add up those disparate features (melancholy, pessimism, skepticism, and so on) that together form the whole quietist *Weltanschauung* from which Beckett's writings emerge.

By the time of his aesthetic essay on James Joyce's "luminaries" in 1929, "Dante...Bruno.Vico..Joyce," Beckett had already reached an uneasy truce with existence by way of, for one, *The Divine Comedy.* In a rare interjection of personal views in his first published essay, Beckett depicts the stasis of Heaven and Hell as no longer a viable artistic subject matter: "On this earth that is Purgatory, Vice and Virtue—which you may take to mean any pair of large contrary human factors—must in turn be purged down to spirits of rebelliousness."[7] Instead, modernist emphasis upon the ongoing purgatorial processes of daily life becomes worthy of celebration in Joyce, and later Marcel Proust (in the 1931 *Proust*). Nevertheless, the penitential quality Beckett found in Dante Aligheri—especially the image of Belacqua, "clasping his knees and holding his face low down between them"—was codified in these early years, as his "Dante notebook" and the creation of an early anti-hero, Belacqua Shuah, make clear.[8] Similarly, Knowlson recounts Beckett's interest in *Unanimisme,* especially the poetry of Jules Romains and Pierre-Jean Jouve, during his final year at TCD: "An outlook that sees the individual as finding some degree of solace in a collective must have held some attraction for a young man who at the time was feeling increasingly his own sense of isolation."[9]

Should Beckett's other literary interests at this time offer any barometer, however, these were much less concerned with intellectual community than with what he calls, in connection with John Keats, "that crouching brood-ing quality."[10] At school with Geoffrey Thompson, Beckett had memorized "Ode to a Nightingale," later citing a line from its sixth stanza—"take into the air my quiet breath"—as fitting evidence of his world-weariness. And in praise of Keats in a letter to MacGreevy from 1931, Beckett noted the latter's downplayed angst: "he doesn't beat his fists on the table."[11] The effect the quoted line, in particular, had upon Beckett, is beautifully woven into the conclusion of "Dante and the Lobster," offering an excellent comment upon the ineluctable nature of suffering: life is *not* a quick death.[12] And the same, understated gloominess pervading Keats' poem was specifically quoted by Beckett, mere weeks before his death, during a visit by his friend, the Irish poet, John Montague.[13] For Beckett, then, the "Ode to a Nightingale" long represented for a trope of artistic mournfulness.

Strangely, the feeling of being "done" communicated with "resignation and, perhaps, disappointment" to Montague in 1989 also elicits shades of Giacomo Leopardi: "Death is not an evil; for it frees man from all evils and, in taking away joy, also removes desire. Old age is the greatest of evils, for it deprives man of every pleasure, while leaving every appetite, and brings with it all sorrows. Yet men fear death, and desire old age."[14] Like Keats, who Beckett also encountered at Trinity College, Dublin, Leopardi's tem-perament may be said to run counter to the positivism of much nineteenth century poetry and prose. Yet clearly Leopardi's "brooding" is much more overt than Keats', and rightly turns our focus from style and understate-ment toward content and sentiment. Indeed, as Ackerley's text on quietism indicates, the thread linking these views is strongly rooted in a personal asceticism, one eschewing the distractions and sufferings of the world as both ceaseless and superfluous.[15] And for Beckett, art offered both the best palliative for, and most effective endorsement of, just such reflections.

Not surprisingly, then, a typed copy of Leopardi's "A Se Stesso" is included in Samuel Beckett's "Interwar Notes."[16] Its importance to Beckett certainly extends to *Proust:* his simile for the world, below, had initially been intended as an epigraph, and Leopardi is marshaled as a "sage" for "wisdom that consists not in the satisfaction but in the ablation of desire."[17] Leopardi's conviction stands as a central buttress to Beckett's quietism, later bequeathed by Schopenhauer: "in noi di cari inganni / non che la speme, il desiderio e spento."[18] The vigor, the uncompromising resentment, and especially the rejection of willing in "A Se Stesso" may together explain why Beckett transcribed Leopardi's "To Himself," quite possibly around the time of writing *Proust:*

> Rest still for ever. You
> Have beaten long enough. And to no purpose

Were all your stirrings; earth not worth your sighs.
Boredom and bitterness
Is life; and the rest, nothing; the world is dirt.[19]

This middle third reproduced in translation here again raises similarity of temperament, while also acting as a kind of biographical mantra for calming the heart—"that cracked beater" subjecting Beckett to panic attacks at this time.[20] Again, the importance of reading "agnostic quietism" into Beckett's life and works is precisely because it is biographically apposite: Beckett felt it himself, strengthened it through his readings in the 1930s, and incorporated a personal shaping of this doctrine into his writings thereafter.

John Pilling's *Beckett's Dream Notebook* is also indispensable in documenting the development of Beckett's quietist outlook. Not surprisingly, *The Imitation of Christ* and *The Confessions* feature prominently, with roughly three dozen entries from the first, and nearly 150 entries from the second. As to the latter, aside from "phrase-hunting" in St. Augustine (entries 354–430),[21] Beckett might also be said to have been "temperament-hunting," as suggested by entries like, "What more miserable than a miserable being who commiserates not himself," and "with a new grief I grieved for my grief and was thus worn by a double sorrow (death of Monica)."[22] That entries like these are ticked and incorporated into *Dream of Fair to Middling Women* reinforces not only the importance of religious texts in Beckett's eclectic reading, but also the forlorn sentiments Beckett emphasized in transcriptions and artistic transformations within such books. Consequently, *Beckett's Dream Notebook* answers important questions raised in Beckett's first sustained piece of writing. The narrator's "Who said all that?," sandwiched between those Augustinian quotations above, thus helps deepen Belacqua's melancholic view of love in the novel, contributing as it does to the mournful conviction that this is a "Beschissenes Dasein beschissenes Dasein" [shitty existence shitty existence]: "But right enough all the same what more miserable than the miserable man that commiserates not himself, caesura, with new grief grieves not for his grief, is not worn by a double sorrow, drowns not in the ken of shore? [. . .] turned he hath and turned again, on back, sides and belly."[23]

Another source highlighted in *Beckett's Dream Notebook* is Robert Burton's *Anatomy of Melancholy*, clearly also important at this time. And comprising as it does fully one-quarter of *University of Reading MS 5000*, is a vital source too frequently overlooked by scholars. Sadly, considerations of space mean much the same injustice will befall Burton's masterpiece here, though we should note that, as with Beckett's 20,000 words of typewritten "Psychology Notes,"[24] an overriding impetus in reading its entirety may have been self-diagnosis. Burton's (and Beckett's) championing of Democritus (in his preface, "Democritus to the Reader," Burton calls himself "Democritus Junior") is also worthy of note in this connection.[25] Though casting himself as a mere shadow of the Abderite, Burton sees himself as carrying on his work, "laughing at the world, consigning himself to libraries, and studying "melancholy and madness [...] so as to better cure it in himself."[26] And herein rests the difference between Burton and Beckett: while Burton's three books comprising the *Anatomy* generally construe melancholy as a condition to be overcome, Beckett's much more artistic construction of melancholy suggests an outlook revealing signs of genius, imagination and a deeper understanding of the "true" nature of existence. In this, it is closer in spirit to Albrecht Dürer's *Melancholia I,* as Giuseppina Restivo has recently pointed out.[27] Artistic melancholy is not a subject of any great interest for Burton, which perhaps explains why entries in *Beckett's Dream Notebook* reveal a "phrase-hunting" similar to that of Augustine's *Confessions:* the sad, purgatorial and suffering sentiments exist in both texts, but as an instruction against despair in reaching toward happiness, instead of creatively harnessing melancholy. For Burton, solitary and idle preconditions should be avoided as "fear and sorrow are the true characters and inseparable companions of most melancholy." For Beckett, in contrast, artists like Keats, Leopardi and Dürer understand melancholy in terms of corporeal asceticism and visionary insight; in short, a ladder toward the "baroque solipsism" mentioned to MacGreevy.[28]

Moreover, the distinction between melancholy as condition and melancholy as temperament may well explain why the Burton entries employed in *Dream of Fair to Middling Women* are generally used, in Dirk van Hulle's phrase, in "the non-syntagmatic way in which Joyce jotted down words in his notebooks," not as lengthy quotations or conceptual borrowings.[29] The manner in which Beckett understood Burton's "causes" of melancholy as desirable, lends a particularly good perspective on the genealogy of "agnostic quietism." For "Democritus Junior" is fundamentally a philosophical optimist ("Again & again I request you to be merry"[30]) writing on melancholy, while Beckett, and to a lesser extent, Belacqua, is a poet finding melancholy more a Pyrrhic facilitation than an impairment for viewing what Burton termed "sky the fairest part of creation":

> When Belacqua came out [...] no moon was to be seen nor stars of any kind. There was no light in the sky whatsoever. At least he could not discover any [...] There was some light, of course there was, it being well known that perfect black is simply not to be had. But he was in no state of mind to be concerned with any such *punctilio.* The heavens, he said to himself, are darkened, absolutely, beyond any possibility of error.[31]

A final example highlighting the importance of *Beckett's Dream Notebook* in the development of "agnostic quietism" is testified by entries 681 and 683—"Pleroma (totality of divine attributes)" and "hypostatized Abstraction."[32] These are taken from W.R. Inge's 1899 *Christian Mysticism,* and are both used in the 1935 letter to MacGreevy to mark Beckett's distance from Christianity, with the former recurring in *Dream of Fair to Middling Women.*[33] More generally, in verifying that Beckett has read this whole book, Pilling's research allows us to apprehend the depth of Beckett's understanding of the term "quietism" used when speaking of MacGreevy's poetry, and later, himself. Inge's work informed Beckett that Miguel de Molinos (c. 1640–1695) initiated "Quietism" in the interests of self-perfection and knowledge of God, which "consists in complete resignation to the will of God, annihilation of all self-will, and an unruffled tranquility or passivity of soul"; and furthermore, "The best kind of prayer is the prayer of silence; and there are three silences, that of words, that of desires, and that of thought. In the last and highest the mind is a blank [...] In this state the soul would willingly even go to hell, if it were God's will."[34] Inge also reveals how Molinos ended his days in a Spanish dungeon for teachings construed as dangerous to Catholicism; thus the satirical link to the "dungeon in Spain" noted in *Murphy* is given added depth: "Stimulated by all those lives immured in mind, as he insisted on supposing, he laboured more diligently than ever before at his own *little dungeon in Spain.*"[35]

Coming to rest here, however, would neglect a central strand of quietism, even as first constructed by Molinos: rejection of the will. Whereas Catholics of Molinos' stamp saw this as a precondition for best communing with God, for one without a "disposition for the supernatural" another sort of guide was needed: a Virgil of melancholy to chart the possibility of will-lessness as an end in itself. Here the most important "quietistic" influence comes via the writings of Schopenhauer, probably first encountered by Beckett in the summer of 1930, just before the writing of *Proust.* Interestingly, the chapter "On the Doctrine of the Denial of the Will-to Live" in *The World as Will and Representation* refers to Molinos and the character of Christian quietism as an ascetic faith expressing Christianity's "deepest truth, its high value, and its sublime character."[36] This is unusually generous praise from Schopenhauer, who typically sees the *practice* of Christianity as antithetical to his philosophy. For this reason, Mark Nixon views Beckett's engagement with this "aesthetic of unhappiness" in terms of "secular quietism."[37] My reading examines Beckett's engagement with quietism in terms of Christian philosophy rather than German literature and culture—and is, I would suggest, more speculative and open-ended (hence "agnostic" rather than "secular"). Yet one must certainly agree with Nixon's claim regarding "Beckett's belief that the solitary state is the irrevocable fate of human beings, particularly pronounced in the artist."[38] What Nixon has clarified is that from the earliest stage of conception, even Beckett's friends knew that the attempted equation would be "Sam. Beckett = Proust + Pessimism."[39] That Beckett later admitted to overstating Proust's pessimism seems due in large measure to his own outlook, forged over the previous 24 years, and the simultaneous reading of Schopenhauer alongside writing *Proust:* "I am going now to buy his 'Aphorisms sur la sagesse dans la vie,' that Proust admired so much for its originality and guarantee of wide reading—transformed. His chapter on 'Will and Representation in Music' is amusing and applies to P.[roust], who certainly read it."[40] In his essay, Beckett describes the influence of Schopenhauer on Proust as "unquestionable,"[41] referring to Book 3, Sect. 52, which makes its way into *Proust* as one of the three explicit references to Schopenhauer: "The influence of Schopenhauer on this aspect of the Proustian demonstration is unquestionable. Schopenhauer rejects the Leibnizian view of music as 'occult arithmetic' and in his aesthetics separates it from the other arts, which can only produce the Idea with its concomitant phenomena, whereas music is the Idea itself."[42] Schopenhauer's views on fine art, from approbations and anathemas to artistic definitions and tropes, undoubtedly contributed to Beckett's views in this area as well. Given Schopenhauer's conviction, well captured by Angela Moorjani, that "art stills temporarily the suffering and terror of existence,"[43] surprisingly little on the marked influence of Schopenhauer has been explored in Beckett studies. Yet those scholars who have examined this angle concur that Beckett had a "sensed affinity" with Schopenhauer, consequently emphasized the latter's pessimism, artistic views and the role of the will; all to the detriment of a more detached, even-handed analysis of Proust's *Remembrance of Things Past.*[44]

Notwithstanding the limited commentary on Beckett's debt to Schopenhauer, Beckett's friend and scholar Gottfried Büttner finds three Schopenhauerian "recommendations" for "enduring the misery of existence": "aesthetic contemplation," noted above, is one. The other two, "compassion and resignation," will be discussed below, using Büttner as a template. Based on his intimacy with Beckett, Büttner is able to link Schopenhauer's pessimism to Beckett's "inner mood," his "melancholic temperament, his inclination to resignation," while simultaneously ensuring such an understanding remains utterly distinct from nihilism.[45] For present purposes, Büttner's final consideration of Schopenhauer's influence is held here to be most striking for the young Beckett: resignation. Just as Beckett's reification of art underscores Büttner's first point, and the religious texts—in particular Kempis—highlighted the importance of compassion, Schopenhauer's unique and profound legacy in Beckett's art is, above all, an acceptance of suffering; a denial of the material world and the desire accompanying it; a jettisoning of hope; or to cite a condition mentioned in *The Unnamable,* a sense of being "lashed to the stake, blindfold, gagged to the gullet."[46]

An excerpt from Schopenhauer beautifully conveys that rejection of the will "recommended" to Beckett, and in turn by Beckett to his creatures:

> The unspeakable pain, the wretchedness and misery of mankind, the triumph of wickedness, the scornful mastery of chance, and the irretrievable fall of the just and the innocent are all here presented to us; and here is to be found a significant hint as to the nature of the world and of existence. It is the antagonism of the will with itself [...] where the phenomenon, the veil of Maya, no longer deceives it. It sees through the form of the phenomenon, the *principium individuationis;* the egoism resting on this expires with it. The *motives* that were previously so powerful now lose their force, and instead of them, the complete knowledge of the real nature of the world, acting as a quieter of the will, produces resignation, the giving up not merely of life, but of the whole will-to-live itself.[47]

Later in this section, Schopenhauer's discussion of tragic literature as the summit of poetical achievement invokes Pedro Calderon's (1600–1681) sin of birth, also explicitly invoked by Beckett in *Proust*. By the second half of 1930, an internalization of Schopenhauer's views on existence become evident in Beckett's writing. Unlike "Dante...Bruno.Vico..Joyce," the demarcation between the start of Beckett's views and the end of those later advanced by his Schopenhauerian Proust is obscured—perhaps purposely so:

> Tragedy is not concerned with human justice. Tragedy is the statement of an expiation, but not the miserable expiation of a codified breach of a local arrangement, organised by the knaves for the fools. The tragic figure represents the expiation of original sin, of the original and eternal sin of him and all his "socii malorum," the sin of having been born.
>
> "Pues el delito mayor
> Del hombre es haber nacido."[48]

Indeed, strange as it sounds, there is no overstatement in asserting the impact of Schopenhauer upon *Proust* to be comparable to the impact of Proust upon *Proust*. Schopenhauer's stamp is affixed to the nostalgically capitalized Idea, Beckett's take on individual perception of the world as a projection of consciousness (and perhaps, indirectly, Beckett's discussion of Time upon Habit as a succession of perceptions made familiar through Memory); and of course, the application of the will-to-live. It is fitting, then, that the final word in *Proust* is Schopenhauer's: "defunctus."

Of course, Schopenhauerian wisdom is not a sufficient condition for the elaboration of an "agnostic quietism." A conflation of factors was necessary in order to construct such an outlook, including the rejection of eternal salvation as envisioned by the Christians and a view of existence as more desolate and punitive than is acceptable to artistic, theological and philosophical optimism. For Beckett, achievements within the arts—as vehicles for the reflection necessary to apprehend, redeem and palliate painful human circumstances—become revelatory. Indeed, art acts as both a melancholic insight into the "true" human condition and a practical mechanism for ameliorating the suffering inherent in that condition. A guarding against willfulness—that "ablation of desire"[49] commended in *Proust*—is another trope, tying together otherwise divergent figures like Leopardi and Molinos, finding an ethics in the hopelessness of struggle and the consequent jettisoning of the will. All of these ideas are restated with great vehemence in Schopenhauer's philosophy, one Beckett admired before writing any of his extended fiction. Aside from the unmistakable contributions by Burton, Augustine, Keats, et al., it is Schopenhauer who appears most influential in giving shape and form to Beckett's construction of "agnostic quietism."

For this reason, the influence of Schopenhauer's ideas upon Beckett demands further consideration, especially his division between individual perception—called by Schopenhauer the *principium individuationis* ("principle of individuation")—and Reality, the thing-in-itself, or what Beckett dubs "non-anthropomorphised humanity" in a letter to MacGreevy: in fine, the world independent of subjectivity.[50] Schopenhauer is absolutely clear on the importance of this point and returns to it again and again. "The veil of Maya" is Schopenhauer's Hindu shorthand for the paradox of humankind expressed through individual existence: "precisely this visible world in which we are, a magic effect called into being, an unstable and inconstant illusion without substance, comparable to the optical illusion and the dream, a veil enveloping human consciousness, a something of which it is equally false and equally true to say that it is and that it is not."[51] More striking still is the importance given to this screen, Beckett's literary "caesura," by Schopenhauer. For the "veil of Maya" is the cause of individual anguish through self-serving pursuits; the impediment to that "deliverance from life and suffering [which] cannot even be imagined without complete denial of the will" existing beyond the "veil of Maya"; and finally and most pivotally, it is the doorway to true compassion itself.[52] Therefore, seeing the "veil of Maya" for what it is—an egoistic delusion, a web of self-interest separating us from other beings—leads both to a rejection of needing to live, an identification with all suffering, and an expression of simultaneous ethical resignation and altruism that Schopenhauer asserts is the summit of human achievement:

> If that veil of Maya, the *prinicipium individuationis,* is lifted from the eyes of a man to such an extent that he no longer makes the egoistical distinction between himself and the person of others, but takes as much interest in the sufferings of other individuals as in his own [...] then it follows automatically that such a man, recognizing in all beings his own true and innermost self, must also regard the endless sufferings of all that lives as his own, and thus take upon himself the pain of the whole world [...] He knows the whole, comprehends its inner nature, and finds it involved in a constant passing away, a vain striving, an inward conflict, and a continual suffering.[53]

These sentiments are critical with regard to Beckett's writing. First, his characters become less and less distinct, the setting more and more timeless, even placeless, in order to approach "the endless sufferings of all that lives"; and secondly, Schopenhauer's ocular metaphor for seeing the world as it is resonates across Beckett's work and addresses many of the same considerations.

This "unveiling" is exemplified by what is called "the vision at last" in *Krapp's Last Tape:* a realization mixing light and dark, alongside a quieting of the will and an acceptance of the human curse of enduring and expressing pain; altogether a worthy and heretofore unexplored artistic seam for Krapp (and Beckett).[54] And for Beckett, the subject of that artistic contemplation may very well be the "veil of Maya" itself. At least, it reflects a reading and artistic interest running from *Proust* and *Beckett's Dream Notebook* to the journalism of the period and, especially, images cast in *Dream of Fair to Middling Women:* "just behind the wall-paper, slashing the close invisible plane with ghastily muted slithers and somersaults. He thought of the rank dark room, quiet, *quieted,* when he would enter, then the first stir behind the paper, the first discreet slithers."[55] Whether Belacqua's wall-paper, Watt's fence with Sam, the curtain in *Ill Seen Ill Said,* or even Murphy's chair—a division between the world and the "close invisible plane" behind it, a trope of veiled perception—can be profitably construed as the mechanism whereby Beckett's "agnostic quietism" becomes a subject for literature. Again and again, through all the affirmations and negations, the creative destructions, Beckett's writing approaches and encroaches the boundaries of the possible, knowable, expressive, real: "I cannot say I am, I can't say anything, I've tried, I'm trying, he knows nothing, knows of nothing, neither what it is to speak, not what it is to hear, to know nothing, to be capable of nothing, and to have to try, you don't try any more, no need to try, it goes on by itself, from word to word [...]."[56]

Lest we place too much importance on the "veil of Maya," two points bear emphasizing. First, even Beckett's obvious indebtedness to Schopenhauer may itself be subsumed into a greater affinity with "artistic melancholy," to employ Giuseppina Restivo's useful phrase. In rightly linking Beckett to Walther von der Vogelweide and Dürer, Restivo reinforces Büttner's Schopenhauerian "recommendations."[57] Both scholars have also been drawn upon in exploring "agnostic quietism," shorthand for both the angst and philosophical inspiration contributing to Beckett's specific experiences and texts during the 1930s and thereafter. And, as with so much of Beckett's reading, what initially appears exhaustive frequently turns out instead to be focused or synoptic. In keeping with his opportunistic use of secondary sources, all the references to Schopenhauer in *Proust* come from Sections 51 (Calderón) and 52 (Leibniz) in *The World as Will and Representation*. These hundred or so pages offer significant discussion on the partition between subjective experience and objective existence, and may also be responsible for Beckett's introduction to Goethe, whom Schopenhauer so venerates that he reproduces the final stanza of "Prometheus" as evidence of the "*denial of the will-to-live*" and the "*quieter* of the will" found beyond the "empty mirage and the web of Maya."[58]

Bearing these reservations in mind, Beckett's 11 August 1936 entry from the "Clare Street Notebook" ties together a number of different themes contributing to Beckett's "agnostic quietism":

Victoria Group

There are moments where the veil of hope is finally ripped away and the eyes, suddenly liberated, see *their* world as it is, as it must be. Alas, it does not last long, the perception quickly passes: the eyes can only bear such a merciless light for a short while, the thin skin of hope re-forms and one returns to the world of phenomena.

Hope is the cataract of the spirit that cannot be pierced until it is ripe for decay. Not every cataract ripens: many a human being spends his whole life enveloped in the mist of hope. And even if the cataract can be pierced for a moment it almost always re-forms immediately; and thus it is with hope. And people never tire of applying to themselves the comforting clichés inspired by hope: hope is the first precondition of life, the instinct that the human race has to thank for not dying out long ago. To thank! Should we really assume as a basic premise that life is so completely unbearable with self-knowledge, that steady, clear self-knowledge whose voice serenely asserts "This is how you are, this is how you will remain. As you have fared until now, so you will continue to fare, till your 'I' decomposes into the parts that are so familiar to you. For you need expect from death nothing either better or worse than this division."[59]

Mark Nixon rightly links this passage to Beckett's contemporaneous reading of Goethe, finding that "this was the zone, the painful but true reality behind the veil, which [Beckett's] writing needed to penetrate." He also suggests the usefulness of a comparison of related metaphors from the famous "German Letter" to Axel Kaun, such as "dissolve" and "porous," and the precise same imagery—in this case "mist"—is raised in personal terms to Axel Kaun a year later: "more and more my own language appears to me like a veil that must be torn apart in order to get at the things (or the Nothingness) behind it."[60] But just as surely as it links the so-called "Victoria Group" in *Faust* to the "veil of Maya" in the *World as Will and Representation,* the excerpt also ties Beckett's own progress from the rather naked pessimism of *Proust,* to the more stoic resignation of the will in his postwar writing. When taken as indicative of Beckett's aesthetic and indeed ethical outlook, the passage suggests a creative approach still being honed along melancholic, skeptical, and nominalist lines.

Even a cursory examination of Beckett's work suggests the importance of this concept. Schopenhauer's description of the sundered "veil of Maya" and the "dread" entailed surface repeatedly. In *Watt,* for example, Watt experiences Schopenhauer's ablation of the will lying beyond the "veil," passing through the hallucinatory doorway of Maya and embracing the Schopenhauerian denial of the will-to-live. The novel concludes with the triumph of "agnostic quietism" over "sufficient reason," the phenomenal world and the will-to-live. Later on, Beckett casts the "veil of Maya"—which Schopenhauer defined as "a something of which it is equally false and equally true to say that it is and that it is not"—as subject matter *per se,* as both subject and object of literature itself. The division between the knowable and the illusory, in this sense, moves beyond Schopenhauerian tropes to become the cause of despair. In this regard, perhaps Beckett's pessimism trumps Schopenhauer's: the latter can find solace in seeing the Thing-In-Itself beyond a deceitful existence; but for Beckett, all is *Ill Seen Ill Said,* veiled or otherwise.

In this way, Beckett may be said to have solved—insofar as the relationship of form and content is soluble—those problems in the "no-man's-land" of subject and object so perplexing him over the 1930s.[61] *Ill Seen Ill*

Said offers a fitting, final perspective upon this change. Right at the start, a black veil separating an old woman's cabin from existence outside is introduced. This curtain, "trembling imperceptibly without cease," is attached to a hook and hung by a nail covering the window: "Opened by her to let her see the sky. But even without that she is there. Without the curtain's being opened. Suddenly open. A flash. The suddenness of it all! She still without stopping. On her way without starting. Gone without going. Back without returning."[62] This passage is immediately followed by the seeming juxtaposition with "the madhouse of the skull and nowhere else." Yet conceiving skull and veil separately, as both dividing objective reality and subjective perception, is tempting but hasty:

> On resumption the head is covered. No matter. No matter now. Such the confusion now between real and—how say its contrary? No matter. That old tandem. Such now the confusion between them once so twain. And such the farrago from eye to mind. For it to make what sad sense of it may. No matter now. Such equal liars both. Real and—how ill say its contrary? The counter-poison.[63]

The identification of the two is made explicit toward the end of *Ill Seen Ill Said*—"Black night fallen. But no. In her head too pure wait"—and thereafter becomes a subject upon which all comes to depend: "But first the partition. It rid they too would be. It less they by as much. It of all the properties doubtless the least obdurate. See the instant see it again when unaided it dissolved."[64] After the dissolution of the veil, what remains is not so much character, setting, even situation, but the "uncommon common noun collapsion": "Then far from the still agonizing eye a gleam of hope. By the grace of these modest beginnings. Within second sight, the shack of ruins. To scrute together with the inscrutable face. All curiosity spent." And with the renunciation of curiosity and desire, what we have here termed "agnostic quietism" is all that remains, alongside the Schopenhauerian paradox—now extending to both concept ("First last moment") and language ("Know happiness")—used to describe the "veil of Maya":

> For the last time at last for to end yet again what the wrong word? Than revoked. No but slowly dispelled a little very little like the last wisps of day when the curtain closes. Of itself by slow millimetres or drawn by phantom hand. Farewell to farewell. Then in that perfect dark foreknell darling sound pip for end begun. First last moment. Grant only enough remain to devour all. Moment by glutton moment. Sky earth the whole kit and boodle. Not another crumb of carrion left. Lick chops and basta.

No. One moment more. One last. Grace to breathe that void. Know happiness.[65]

In creating this "Beckettian" vision, three especially formative influences have been considered in the development of Beckett's "agnostic quietism" in the 1930s: an artistic trope of melancholic contemplation; an appreciation of human suffering in a nasty world and the stoicism and resignation necessary to go on despite the anguish entailed therein; and finally, the ethical and aesthetic acceptance of a veil separating perception from truth, hope from will-lessness. The third aspect not only becomes the basis for consolation—without hope of Christian salvation, what remains is a shared community of pain mandating compassion for others—but also, I suggest, this Schopenhauerian partition increasingly becomes a literary subject matter itself, situated as "it" is between Beckett's concern with the "no-man's-land" dividing subject and object. In many ways, Schopenhauer's "recommendations" were paradigmatic in Beckett's emerging worldview, yet these were not exclusive: as Beckett often suggested, his tools were impotence and ignorance. Both were ancillary concerns for Schopenhauer, since a paradoxical strength and knowledge arises from denial of the will-to-live, but for Beckett these became the foundation stones of his art. Here a bulk of archival materials support the view that Beckett—already well on the way to developing his own personal and artistic ethos—became interested in that infamous incapacity of deed and word so pervasive in his later works, one achieved through an intellectual mosaic patched together from the interwar readings and note-taking just beginning to be mined by Beckett Studies.

Notes

1. Samuel Beckett-Thomas MacGreevy Correspondence, Trinity College Dublin, MS 10904; © Samuel Beckett Estate. English translations of the original Latin can be found in John Pilling's *Beckett's Dream Notebook* (Reading: Beckett International Foundation, 1999), 85–86; further excerpts from this self-reflexive letter are contained in James Knowlson's *Damned to Fame: The Life of Samuel Beckett* (London: Bloomsbury, 1996), 179–81. I am indebted to Mark Nixon and Erik Tonning for their assistance with portions of this letter.
2. See Mark Nixon, "'Scraps of German': Beckett Reading German Literature," in Matthijs Engelberts, Everett Frost, with Jane Maxwell, ed., "Notes diverse holo," *Samuel Beckett Today/Aujourd'hui* 16 (Amsterdam: Rodopi, 2006), 264, 278; Knowlson, *Damned to Fame,* 353; C.J. Ackerley, "Samuel Beckett and Thomas á Kempis: The Roots of Quietism," in Marius Buning, Matthijs Engelberts, Onno Kosters, eds., "Beckett and Religion/Beckett/

Aesthetics/Politics / Beckett et la religion/Beckett/L'Esthétique/La politique," *Samuel Beckett Today/Aujourd'hui* 9 (Amsterdam: Rodopi, 2000), 81–92.

3. Ibid., 88. It is however, worth noting that Kempis' fifteenth century text was not intended as mysticism of any kind (though it reads that way now)—one reason it was never placed on the codex of banned books during the Catholic Counter Reformation a century later—in fact, it was intended to be a sort of guidebook for monks and monastic aspirants.

4. Beckett, Beckett/MacGreevy letters, 10 March 1935.

5. Samuel Beckett, "Humanistic Quietism," in *Disjecta: Miscellaneous Writings and a Dramatic Fragment,* ed. Ruby Cohn (London: Calder, 1983), 68.

6. Ibid., 68.

7. Samuel Beckett, "Dante... Bruno. Vico.. Joyce," in *Disjecta,* 33.

8. Dante, *The Divine Comedy of Dante Alighieri; II: Purgatory,* Canto IV, 61; Samuel Beckett, "Dante Notebook," Trinity College, Dublin MS 10966/1. © Samuel Beckett Estate.

9. Knowlson, *Damned to Fame,* 76.

10. Beckett, Beckett/MacGreevy undated (probably Spring 1930) letter. © Samuel Beckett Estate.

11. Knowlson, *Damned to Fame,* 117.

12. Samuel Beckett, *More Pricks than Kicks* (London: Calder, 1993), 21.

13. John Montague, "A Few Drinks and a Hymn: My Farewell to Samuel Beckett," *The New York Times,* Sunday Late Edition, April 17, 1994, http://www.samuel-beckett.net/beckett_hymn.html.

14. Giacomo Leopardi, *Selected Prose and Poetry,* trans. I. Origo and J. Heath-Stubbs (London: Oxford University Press, 1966), 190.

15. Ackerley, "Roots of Quietism," 81–82.

16. For details of these notebooks, see my *Beckett's Books: A Cultural History of Samuel Beckett's "Interwar Notes"* (London: Continuum, 2006).

17. Samuel Beckett, *Proust and Three Dialogues with Georges Duthuit* (London: Calder, 1993), 18. For Beckett's intended use of Leopardi as an epigraph in *Proust,* see J.D. O'Hara's "Beckett's Schopenhauerian Reading of Proust: the Will as Whirled in Re-Presentation," in *Schopenhauer: New Essays in Honor of His 200th Birthday,* ed. Eric van der Luft (Lewiston: The Edwin Mellen Press, 1988), 276.

18. "[Not only the dear hope / Of being deluded gone, but the desire]"—a phrase pared down and used in *Molloy.* Samuel Beckett, *The Beckett Trilogy* (London: Picador, 1979), 34.

19. Transcript of "A Se Stesso," Trinity College, Dublin MS 10971/9. © Samuel Beckett Estate.

20. The reference is to *Whoroscope.* Samuel Beckett, *Collected Poems,* 2 (London: Calder, 1999). See also Knowlson, *Damned to Fame,* 174–81.

21. There are 133 hunted phrases on Augustine's *Confessions;* see Pilling, *Beckett's Dream Notebook,* 11–30.

22. Ibid., 12, 17, 25.

23. Samuel Beckett, *Dream of Fair to Middling Women* (Dublin: Black Cat Press, 1992), 72–74.

24. Matthew Feldman, "Beckett's Poss and the Dog's Dinner: An Empirical Survey of the 1930s 'Psychology' and 'Philosophy Notes,'" *Journal of Beckett Studies* 13.2 (2004): 69–94.
25. Robert Burton, *Anatomy of Melancholy,* vol. I, ed. A.R. Shilleto (London: George Bell and Sons, 1904), 16.
26. Ibid., 20.
27. Giuseppina Restivo, "*Melencolias* and Scientific Ironies in *Endgame:* Beckett, Walther, Dürer, Musil," in Angela Moorjani et al., eds., "Endlessness in the Year 2000/Fin sans Fin en L'an 2000," *Samuel Beckett Today/Aujourd'hui* 11 (Amsterdam: Rodopi, 2002), 103–13.
28. Knowlson, *Damned to Fame,* 117.
29. "'Nichtsnichtsundnichts': Beckett's and Joyce's Transtextual Undoings," in *Beckett, Joyce and the Art of the Negative, European Joyce Studies* 16, ed. Colleen Jaurretche (Amsterdam: Rodopi, 2005), 54.
30. Pilling, *Beckett's Dream Notebook,* 114.
31. Burton, *Anatomy of Melancholy,* vol. II, 46; Beckett, *Dream of Fair to Middling Women,* 240–241.
32. Ibid., 98–99.
33. Beckett, *Dream of Fair to Middling Women,* 42.
34. William Inge, *Christian Mysticism: The Bampton Lectures* (London: Metheun & Co, 1921), 232–33.
35. Samuel Beckett, *Murphy,* 100–102, my italics. Further, as an ex-theology student, Murphy's self-bondage in the chair ironically contrasts with Molinos' imposed fate, given their shared rejection of the world—though one suspects their reasoning might be different. The above is but one example of the "what" in which a quietistic reading of *Murphy,* even if strictly along Inge's interpretation, can offer interesting insights. This, in turn, corresponds to entries on Ernest Jones' *Treatment of the Neuroses*—"Dungeons in Spain. (Mine own.)"—in Beckett's "Psychology Notes." Trinity College, Dublin MS 10971/8/21. © Samuel Beckett Estate.
36. Arthur Schopenhauer, *The World as Will and Representation,* vol. II, trans. E.F.J. Payne (New York: Dover, 1969), 616.
37. Mark Nixon, "'what a tourist I must have been': The German Dairies of Samuel Beckett," (Unpublished Ph.D. Thesis: University of Reading, 2005), 40, 63.
38. Ibid., 58.
39. Entry in George Reavey's diary, 15 July 1930, quoted in Nixon, "Scraps of German," 278–79.
40. I am most grateful to Mark Nixon and John Pilling for their assistance with this letter.
41. Beckett, *Proust,* 91–92.
42. Ibid. The other references to "an objectivation of the individual's will, Schopenhauer would say" and Schopenhauer's definition of the artistic procedure as "the contemplation of the world independently of the principle of reason," are found on 19 and 87.

43. Angela Moorjani, "Mourning, Schopenhauer, and Beckett's Art of Shadows," in *Beckett On and On,* ed. Lois Oppenheim and Marius Buning (London: Fairleigh Dickenson University Press, 1996), 85.
44. O'Hara, "Beckett's Schopenhauerian Reading of Proust," 275, 285; Steven J. Rosen, *Samuel Beckett and the Pessimistic Tradition* (New Brunswick: Rutgers University Press, 1976), 152.
45. Gottfried Büttner, *Samuel Beckett's Novel Watt,* trans. Joseph P. Dolan (Philadelphia: University of Pennsylvania Press, 1984), 114–15. He quotes Beckett's own rejection of the label: "I simply cannot understand why some people call me a nihilist. There is not basis for that," 122.
46. Beckett, *The Beckett Trilogy,* 360.
47. Schopenhauer, *The World as Will and Representation,* vol. I, 253.
48. Beckett, *Proust,* 67. Pilling's "Beckett's *Proust*" has also shown that quotation of Calderón's maxim—translated as "For man's greatest offence / Is that he has been born," comes via Sect. 51 of the *World as Will and Representation,* 12.
49. Beckett, *Proust,* 18.
50. Beckett, Beckett/MacGreevy letters, 31 January 1938. © Samuel Beckett Estate.
51. Schopenhauer, *World as Will and Representation,* vol. I, 419.
52. Schopenhauer, *World as Will and Representation,* vol. I, 352, 397.
53. Ibid., 378.
54. Samuel Beckett, *Krapp's Last Tape,* in *The Complete Dramatic Works* (London: Faber and Faber, 1990), 220.
55. Beckett, *Dream of Fair to Middling Women,* 15.
56. Beckett, *The Beckett Trilogy,* 370.
57. Beckett's notes on Dürer can be found in UoR MS 5001. Beckett's notes on Vogelweide come from J.G. Robertson's *History of German Literature* which Nixon's "scraps of German," dates to 1934, 262–63.
58. Schopenhauer, *The World as Will and Representation,* vol. I, 285, 284. Beckett also reproduced a typescript copy of Faust's "Prometheus" in German in Trinity College, Dublin MS 10971/1/72. © Samuel Beckett Estate.
59. UoR MS 5003, 33, 35, 37. © Samuel Beckett Estate. For the entire entry, see my "Sourcing 'Aporetics': An Empirical Study on Philosophical Influences in the Development of Samuel Beckett's Writing," (Unpublished Ph.D. Thesis, Oxford Brookes University, 2004), 394–95. I am especially grateful to Mark Nixon for alerting me to this passage, in addition to our lengthy discussions on this important matter.
60. Samuel Beckett, "German Letter of 1937," in *Disjecta,* 170–173; Nixon, "scraps of German," 273–74; Nixon "what a tourist," 178–79.
61. Samuel Beckett, "Recent Irish Poetry," in *Disjecta,* 70.
62. Samuel Beckett, *Ill Seen Ill Said* (London: Calder, 1982), 15, 18–19.
63. Ibid., 40.
64. Ibid., 47, 53.
65. Ibid., 55, 59.

Bibliography

à Kempis, Thomas. *The Imitation of Christ*. Translated by Bernard Bangley. Guildford: Highland Books, 1983.

Abbott, H. Porter. *Beckett Writing Beckett: The Author in the Autograph*. Ithaca and London: Cornell University Press, 1996.

———. "Beginning Again: The Post-narrative Art of *Texts for Nothing* and *How it is*." In *The Cambridge Companion to Beckett*. Edited by John Pilling. Cambridge: Cambridge University Press, 1994. 106–23.

Abraham, Nicholas and Maria Torok. *The Wolf Man's Magic Word: A Cryptonymy*. Translated by Nicholas Rand. Minneapolis: University of Minnesota Press, 1986.

Acheson, James. *Samuel Beckett's Artistic Theory and Practice: Criticism, Drama, and Early Fiction*. Basingstoke: Macmillan, 1997.

Ackerley, C.J. "Samuel Beckett and Max Nordau: Degeneration, Sausage Poisoning, the Bloody Rafflesia, Coenaesthesis, and the Not-I." In *Beckett after Beckett*. Edited by S.E. Gontarski and C.J. Ackerley. Gainesville: University Press of Florida, 2006. 167–76.

Ackerley, Chris. "Samuel Beckett and Thomas á Kempis: The Roots of Quietism." In "Beckett and Religion/Beckett/Aesthetics/Politics / Beckett et la religion/ Beckett/L'Esthétique/La politique." *Samuel Beckett Today/Aujourd'hui* 9. Edited by Marius Buning, Matthijs Engelberts, and Onno Kosters. Amsterdam: Rodopi, 2000. 81–92.

Ackerley, C.J. and S.E. Gontarski. *The Grove Companion to Samuel Beckett*. New York: Grove Press, 2004.

Adelman, Gary. *Naming Beckett's Unnamable*. Lewisburg: Bucknell University Press, 2004.

Adorno, Theodor W. Inscribed Typescript of *Endgame* Lecture. *Sotheby's: English Literature*. Auction Catalogue. London: Sotheby's, July 8, 2004.

———. "Trying to Understand *Endgame*." In *Can One Live After Auschwitz? A Philosophical Reader*. Edited by Rolf Tiedemann. Stanford: Stanford University Press, 2003. 259–95.

———. *The Adorno Reader*. Edited by Brian O'Connor. Oxford: Blackwell, 2000.

Adorno, Theodor W. *Aesthetic Theory*. Edited and translated by Robert Hullot-Kentor. Minneapolis: University of Minnesota Press, 1997.

———. *Prisms*. Translated by Samuel Weber and Sherry Weber. Cambridge, MA: MIT Press, 1981.

———. *The Jargon of Authenticity*. Evanston: Northwestern University Press, 1973.

———. *Negative Dialectics*. Translated by E.B. Ashton. London: Routledge & Kegan Paul, 1973.

———. "Towards an Understanding of *Endgame*." In *Twentieth Century Interpretations of Endgame*. Edited by Bell Gale Chevigny. Translated by Samuel Weber. New Jersey: Prentice-Hall, 1969. 82–114.

Agamben, Giorgio. *Profanations*. Translated by Jeff Fort. New York: Zone Books, 2007.

———. *Means Without Ends: Notes on Politics*. Translated by Vincenzo Binetti and Cesare Casarino. Minneapolis: University of Minnesota Press, 2000.

———. *Remnants of Auschwitz: The Witness and the Archive*. New York: Zone Books, 1999.

———. *Homo Sacer: Sovereign Power and Bare Life*. Stanford: Stanford University Press, 1998.

Ahmad, Aijaz. *In Theory: Classes, Nations, Literatures*. London: Verso, 1992.

Albrecht, Klaus. "Günter Albrecht—Samuel Beckett—Axel Kaun." Translated by Mark Nixon. *Journal of Beckett Studies* 13.2 (Spring 2004): 24–38.

Anderson, Benedict. *Imagined Communities: Reflections on the Origin and Spread of Nationalism*. London: Verso, 1983.

Arikha, Avigdor. *Boyhood Drawings Made in Deportation*. Paris: Les Amis de l'Aliya des Jeunes, 1971.

Armstrong, Gordon Scott. *Theatre and Consciousness: The Nature of Bio-Evolutionary in the Arts*. New York: Peter Lang, 2003.

Atik, Anne. *How It Was: A Memoir of Samuel Beckett*. London: Faber & Faber, 2001.

Badiou, Alain. *On Beckett: Dissymetries*. Edited and translated by Nina Power and Alberto Toscano. Manchester: Clinamen Press, 2003.

Bair, Deirdre, *Samuel Beckett: A Biography*. London: Picador, 1978.

———. *Samuel Beckett: A Biography*. New York: Harcourt Brace Jovanovich, 1978.

Banfield, Ann. "Beckett's Tattered Syntax." *Representations* 84 (Autumn 2003): 6–29.

Banville, John. "Words from the Witness." *The Irish Times Weekend Review*. September 25, 2004.

Barthes, Roland. *Mythologies*. New York: Hill and Wang, 1972.

Beckett, Samuel. "Avant *Fin de partie*" Typescript. Beckett International Foundation, UoR MS 1227/7/16/7.

———. The Beckett/Bray Correspondence. Trinity College, Dublin. MS 10948.

———. Charles Prentice from Chatto Letter, 10 November 1933. Beckett International Foundation, UoR.

———. "Ernest et Alice" Typescript. Beckett International Foundation, UoR MS 1227/7/16/2.

———. German Diaries. Beckett International Foundation, UoR.

———. German Vocabulary Notebooks. Beckett International Foundation, UoR MS 5006.

———. Letter to Günter Albrecht, 31 December 1936. Beckett International Foundation, UoR.

———. Letter to Kay Boyle, 12 April 1967. Harry Ransom Center. University of Texas, Austin.

———. "Kilcool" Fragments. Trinity College, Dublin. Holograph notebook TCD MS 4664, ff. 10r–19r.

———. "Last Soliloquy" Manuscript. Beckett International Foundation, UoR MS 2937/1–3.

———. Letter to A.J. Leventhal, 7 May 1934. Harry Ransom Center. University of Texas, Austin.

———. The Mary Manning Howe Correspondence. Harry Ransom Center. University of Texas, Austin.

———. The Susan Manning Correspondence. Harry Ransom Center. University of Texas, Austin.

———. "Notes on Germany, Europe, and the French Revolution." Trinity College, Dublin. MS 10969.

———. "Notes on Painting." Beckett International Foundation, UoR MS 5001.

———. The Thomas MacGreevy Correspondence. Trinity College, Dublin. MS 10402.

———. The George Reavey Correspondence. Harry Ransom Center. University of Texas, Austin.

———. "Sam Francis" Notebook. Beckett International Foundation, UoR MS 2926.

———. "A Se Stesso." Trinity College, Dublin. MS 10971/9.

———. Preparatory Fragments for Stirring Still. Beckett International Foundation. UoR MS 2935/1/2.

———. Letter to Arland Ussher, 6 April 1938. Harry Ransom Center. University of Texas, Austin.

———. Watt Manuscripts. Harry Ransom Center. University of Texas, Austin.

———. Watt Notebook. Harry Ransom Center. University of Texas, Austin.

———. "Whoroscope" Notebook. Beckett International Foundation, UoR MS 3000.

———. Dialogue between X and F. Trinity College, Dublin. MS 4662.

———. Samuel Beckett: Poems, Short Fiction, Criticism. New York: Grove Press, 2006.

———. Premier Amour. Paris: Les Editions de Minuit, 2003.

———. First Love and Other Novellas. Edited by Gerry Dukes. London: Penguin, 2000.

———. Collected Poems, 2. London: Calder, 1999.

———. Proust and Three Dialogues with George Duthuit. London: Calder, 1999.

———. Watt. New York: Grove Press, 1998.

———. How it is. London: Calder, 1996.

Beckett, Samuel. *The Complete Short Prose 1929–1989.* Edited with introduction and notes by S.E. Gontarski. New York: Grove Press, 1995.

———. *More Pricks Than Kicks.* London: Calder, 1993.

———. *Murphy.* London: Calder, 1993.

———. *Dream of Fair to Middling Women.* Dublin: Black Cat Press, 1992.

———. *Watt.* London: Picador, 1988.

———. *The Complete Dramatic Works.* London: Faber & Faber, 1986.

———. *The Collected Shorter Plays of Samuel Beckett.* New York: Grove, 1984.

———. *Disjecta: Miscellaneous Writings and a Dramatic Fragment.* Edited by Ruby Cohn. London: Calder, 1983.

———. *Worstward Ho.* London: Calder, 1983.

———. *The Beckett Trilogy.* London: Picador, 1979.

———. *Stories and Texts for Nothing.* New York: Grove Press, 1967.

———. *Watt.* London: Calder, 1976.

———. *Mercier and Camier.* London: Calder, 1974.

———. *Malone Dies.* London: Penguin, 1965.

———. *Happy Days.* New York: Grove, 1961.

———. *Endgame.* New York: Grove, 1958.

———. *Three Novels: Molloy, Malone Dies, The Unnamable.* New York: Grove Press, 1958.

———. *Proust.* New York: Grove, 1931.

Bellemin-Noël, Jean. *Le texte et l'avant-texte: Les brouillons d'un poème de Milosz.* Paris: Librairie Larousse, 1972.

Beller, Steven. *Antisemitism: A Very Short Introduction.* Oxford: Oxford University Press, 2007.

Benjamin, Walter. *Illuminations.* Edited and translated by Hannah Arendt. New York: Schocken, 1968.

Ben-Zvi, Linda. "*Not I:* Through a Tube Starkly." In *Samuel Beckett.* Edited by Jennifer Birkett and Kate Ince. London and New York: Longman, 2000. 259–65.

Bernstein, Michael André. *Foregone Conclusions: Against Apocalyptic History.* Berkeley: University of California Press, 1994.

Biasi, Pierre-Marc de. *La Génétique des textes.* Paris: Nathan, 2000.

———. "What Is a Literary Draft? Toward a Functional Typology of Genetic Documentation." *Yale French Studies* 89 (1996): 26–58.

Blackman, Jackie. "Postwar Beckett: Resistance, Commitment or Communist Krap?" In *Beckett and Ethics.* Edited by Russell Smith. London: Continuum, forthcoming.

———. "Beckett Judaizing Beckett: 'A Jew from Greenland' in Paris." In "'All Sturm and no Drang': Beckett and Romanticism/Beckett at Reading 2006." *Samuel Beckett Today/Aujourd'hui* 18. Edited by Dirk Van Hulle and Mark Nixon. Amsterdam: Rodopi, 2007. 325–40.

Boulter, Jonathan. "Writing Guilt: Haruki Murakami and the Archives of National Mourning." *English Studies in Canada* 32.1 (March 2006): 125–45.

———. "The Melancholy Archive: Jose Saramago's *All the Names.*" *Genre.* 37.1/2 (Spring/Summer 2005): 115–43.

————. "Does Mourning Require a Subject? Samuel Beckett's *Texts for Nothing.*" *Modern Fiction Studies* 50.2 (Summer 2004): 332–50.

Bourdieu, Pierre. *The Logic of Practice.* Translated by Robert Nice. Stanford: Stanford University Press, 1990.

Bowen, Elizabeth. *The Mulberry Tree.* London: Vintage, 1999.

————. *The Mulberry Tree: Writings of Elizabeth Bowen.* London: Harcourt Brace Jovanovich, 1986.

————. *Seven Winters.* London: Longmans, 1943.

Bowles, Patrick. "A Portfolio and an Interview." *The Paris Review* 33 (Winter–Spring 1965): 46–52.

Boxall, Peter. "Samuel Beckett: Towards a Political Reading." *Irish Studies Review* 10.2 (2002): 159–71.

————. "Introduction to 'Beckett/Aesthetics/Politics.'" In "Beckett and Religion/ Beckett/Aesthetics/Politics." *Samuel Beckett Today/Aujourd'hui* 9. Edited by Marius Buning, Matthijs Engleberts, Onno Rutger Kosters, Mary Bryden, Lance St John Butler, and Peter Boxall. Amsterdam: Rodopi, 2000. 207–9.

Bryden, Mary. *Samuel Beckett and the Idea of God.* New York: St. Martin's Press, 1998.

Bryden, Mary, Julian Garforth, and Peter Mills. *Beckett at Reading: Catalogue of the Beckett Manuscript Collection at the University of Reading.* Reading: Whiteknights Press and the Beckett International Foundation, 1998.

Burton, Robert. *Anatomy of Melancholy.* Vol. I. Edited by A.R. Shilleto. London: George Bell and Sons, 1904.

Bush, Vannevar. "As We May Think." *The Atlantic Monthly.* July 1945, <http:// www.theatlantic.com/unbound/flashbks/computer/bushf.htm>.

Büttner, Gottfried. *Samuel Beckett's Novel* Watt. Translated by Joseph P. Dolan. Philadelphia: University of Pennsylvania Press, 1984.

Callil, Carmen. *Bad Faith: A Forgotten History of Family and Fatherland.* London: Jonathan Cape, 2006.

Caruth, Cathy. *Unclaimed Experience: Trauma, Narrative, and History.* Baltimore and London: Johns Hopkins University Press, 1996.

Casanova, Pascale. *Samuel Beckett: Anatomy of a Literary Revolution.* New York: Verso, 2006.

Casey, Edward. *Getting Back into Place: Towards a Renewed Understanding of the Place-World.* Bloomington: Indiana University Press, 1993.

Channin, Robert and Samuel Beckett. *Arikha.* Paris: Herman, 1985.

Cohn, Ruby. *A Beckett Canon.* Ann Arbor: University of Michigan Press, 2001.

————. *Just Play: Beckett's Theater.* Princeton: Princeton University Press, 1980.

————. *Back to Beckett.* Princeton: Princeton University Press, 1976.

Connor, Steven. *Samuel Beckett: Repetition, Theory and Text.* Oxford: Blackwell, 1988.

Conti, Chris. "Critique and Form: Adorno on *Godot* and *Endgame.*" In "After Beckett/Après Beckett." *Samuel Beckett Today/Aujourd'hui* 14. Edited by Sjef Houppermans, Anthony Uhlmann and Bruno Clement. Amsterdam: Rodopi, 2004. 277–93.

Cronin, Anthony. *Samuel Beckett: The Last Modernist.* New York: Harper Collins, 1997.

Curtis, L.P., Jr. "The Anglo-Irish Predicament." *Twentieth Century Studies* 4 (1970): 46–62.

Dante. *The Divine Comedy of Dante Alighieri; II: Purgatory.* Translated by John D. Sinclair. London: John Lane the Bodley Head, 1948.

Darwin, Charles. *Autobiography.* In *The Life of Charles Darwin,* by Francis Darwin. London: Studio Editions, 1995. 5–54.

Darwin, Francis. *The Life of Charles Darwin.* London: Studio Editions, 1995.

———, ed. *The Foundations of The Origin of Species: Two Essays Written in 1842 and 1844.* Cambridge: Cambridge University Press, 1909.

Deleuze, Gilles and Félix Guattari. *A Thousand Plateaus: Capitalism and Schizophrenia.* Translated by Brian Massumi. London: Athlone Press, 1999.

———. *Mille plateaux: capitalisme et schizophrénie.* Paris: Les éditions de minuit, 1980.

Derrida, Jacques. *The Work of Mourning.* Edited by Pascale-AnneBrault and Michael Naas. Chicago: University of Chicago Press, 2001.

———. *Archive Fever: A Freudian Impression.* Chicago: University of Chicago Press, 1998.

———. *Specters of Marx: The State of the Debt, the Work of Mourning, and the New International.* New York: Routledge, 1994.

———. "Spectres of Marx." *New Left Review* 1.205 (May–June 1994): 31–58.

———. *The Ear of the Other: Otobiography, Transference, Translation.* Edited by Christie McDonald. Lincoln: University of Nebraska Press, 1988.

———. "Foreword: Fors: The Anglish Words of Nicolas Abraham and Maria Torok." *The Wolf Man's Magic Word: A Cryptonymy.* Translated by Nicholas Rand. Minneapolis: University of Minnesota Press, 1986. xi–xlviii.

Dickinson, Page L. *The Dublin of Yesterday.* London: Metheun, 1929.

Diedrich, Antje. "Performance as Rehearsal: George Tabori's Staging of Beckett's *Waiting for Godot* and *Endgame.*" In "Historicising Beckett/Issues of Performance." *Samuel Beckett Today/Aujourd'hui* 15. Edited by Marius Buning, Matthijs Engelberts, Sjef Houppermans, and Dirk Van Hulle. Amsterdam: Rodopi, 2005. 147–60.

Eagleton, Terry. "Beckett and Nothing." In *Reflections on Beckett: A Centenary Tribute.* Edited by Anna McMullan and Steve Wilmer. Ann Arbor: University of Michigan Press, forthcoming 2009.

Earnshaw, Steven. *The Direction of Literary Theory.* London: Macmillan, 1996.

Eng, David L. and David Kazanjian, eds. *Loss: The Politics of Mourning.* Berkeley: University of California Press, 2003.

Enzensberger, Hans Magnus. "Die Entstehung eines Gedichts." In *Gedichte. Die Entstehung eines Gedichts.* Frankfurt am Main: Suhrkamp, 1962. 55–79.

Esslin, Martin. *The Theatre of the Absurd.* 3rd ed. London: Penguin, 1980.

Fehsenfeld, Martha and Dougald McMillan. *Beckett in the Theater.* London: Calder, 1988.

Feinberg, Anat. *Embodied Memory: The Theatre of George Tabori.* Iowa City: University of Iowa Press, 1999.

Feldman, Matthew. *Beckett's Books: A Cultural History of Samuel Beckett's Interwar Notes.* London: Continuum Books, 2006.

———. "Beckett's Poss and the Dog's Dinner: An Empirical Survey of the 1930s 'Psychology' and 'Philosophy Notes.'" *Journal of Beckett Studies* 13.2 (2004): 69–94.

———. "Sourcing 'Aporetics': An Empirical Study on Philosophical Influences in the Development of Samuel Beckett's Writing." Unpublished Ph.D. Thesis, Oxford Brookes University: 2004.

Felman, Shoshana and Dori Laub. *Testimony: Crises of Witnessing in Literature, Psychoanalysis, and History.* New York: Routledge, 1992.

Foucault, Michel. *The Foucault Reader.* Edited by Paul Rabinow. London: Penguin, 1991.

Fraser, Graham. "The Pornographic Imagination in *All Strange Away.*" *Modern Fiction Studies* 41.3–4 (1995): 515–30.

Freud, Sigmund. *The Interpretation of Dreams.* In *The Penguin Freud Library.* Vol. 4. Edited by Angela Richards. London: Penguin. 1991.

———. "Beyond the Pleasure Principle." In *On Metapsychology: The Theory of Psychoanalysis. The Penguin Freud Library.* Vol. 11. Edited by Angela Richards. London: Penguin, 1991. 269–338.

———. "Mourning and Melancholia." In *On Metapsychology: The Theory of Psychoanalysis.* Edited by Angela Richards. Harmondsworth: Penguin, 1991. 251–268.

———. "Mourning and Melancholia." In *The Penguin Freud Library,* Vol. 2. Edited by Angela Richards. London: Penguin. 1984. 245–268.

Gaffney, Phyllis. "Dante, Manzoni, De Valera, Beckett…? Circumlocutions of a Storykeeper: Beckett and Saint-Lô." *Irish University Review* 29.2 (Autumn/ Winter 1999): 256–80.

———. *Healing Amid the Ruins: The Irish Hospital at Saint-Lô.* Dublin: A&A Farmar, 1999.

Gibson, Andrew. "Afterword: the skull the skull the skull the skull in Connemara: Beckett, Ireland and Elsewhere." In *Beckett and Ireland.* Edited by Seán Kennedy. Cambridge: Cambridge University Press, forthcoming.

Gilman, Richard. "*Endgame.*" In *Samuel Beckett's Endgame: Modern Critical Interpretations.* Edited by Harold Bloom. New York: Chelsea House, 1988. 79–86.

———. "Beckett." *Partisan Review* 41 (1974): 56–76.

Gontarski, S.E. *The Intent of Undoing in Samuel Beckett's Dramatic Texts.* Bloomington: Indiana University Press, 1985.

———. "Reinventing Beckett." *Modern Drama* 49.4 (Winter 2006): 419–27.

Gontarski, S.E. and Anthony Uhlmann, eds. "Afterimages: Introducing Beckett's Ghosts." In *Beckett After Beckett.* Tallahassee: University of Florida Press, 2006. 1–12.

Graver, Lawrence and Raymond Federman, eds. *Samuel Beckett: The Critical Heritage.* London: Routledge, 1979.

Grésillon, Almuth. *Eléments de critique génétique: Lire les manuscrits modernes.* Paris: Presses Universitaires de France, 1994.

Gribben, Darren. "Samuel Beckett: Number 465. Censorship of the Self and Imagination in Beckett's Work after World War II." In "Three Dialogues Revisited/Les Trois dialogues revisités." *Samuel Beckett Today/Aujourd'hui* 13. Edited by Matthijs Engelberts, Marius Buning, Danièle de Ruyter Tognotti, and Sjef Houppermans. Amsterdam: Rodopi, 2003. 215–27.

Gruber, Howard E. *Darwin on Man: A Psychological Study of Scientific Creativity.* London: Wildwood House, 1974.

Grubgeld, Elizabeth. "Class, Gender and the Forms of Narrative: The Autobiographies of Anglo-Irish Women." In *Representing Ireland: Gender, Class and Nationality.* Edited by Susan Shaw Sailer. Gainesville: Florida University Press, 1997. 133–55.

———. "Anglo-Irish Autobiography and the Genealogical Mandate." *Eire/Ireland* 32.4 and 33.1 and 2 (1997–1998): 96–115.

Halbwachs, Maurice. *On Collective Memory.* Chicago: University of Chicago Press. 1992.

Hamilton, Elizabeth. *An Irish Childhood.* London: Chatto and Windus, 1963.

Harmon, Maurice, ed. *No Author Better Served: The Correspondence of Samuel Beckett and Alan Schneider.* London: Harvard University Press: 1998.

Hart, Peter. "The Protestant Experience of Revolution in Southern Ireland." In *Unionism in Modern Ireland: New Perspectives on Politics and Culture.* Edited by R. English and G. Walker. London: Macmillan, 1996. 81–98.

Hartman, Geoffrey. "Shoah and the Intellectual Witness." *Partisan Review* 1 (Winter 1998): 37–48.

Hayman, David. "Beckett's *Watt,* the Art-Historical Trace: An Archeological Inquest." *Journal of Beckett Studies* 13.2 (Spring 2004): 95–109.

Heffernan, Teresa. "*Beloved* and the Problem of Mourning." *Studies in the Novel* 30.4 (Winter 1998): 558–573.

Hesla, David. *The Shape of Chaos: An Interpretation of the Art of Samuel Beckett.* Minneapolis: University of Minnesota Press, 1971.

———. "The Shape of Chaos: A Reading of Beckett's *Watt.*" *Critique: Studies in Modern Fiction* 6.1 (1963): 85–105.

Hill, Leslie. "Beckett, Writing, Politics: Answering for Myself." In "Beckett and Religion / Beckett/Aesthetics/Politics." *Samuel Beckett Today/Aujourd'hui* 9. Edited by Marius Buning, Matthijs Engelberts, and Onno Kosters. Amsterdam: Rodopi, 2000. 215–21.

———. "'Up the Republic!': Beckett, Writing, Politics." *MLN* 112 (1997): 909–28.

———. "Samuel Beckett (1906–1989)." In *Radical Philosophy—Summer 1990.* http://www.radicalphilosophy.com/default.asp?channel_id=2191&editorial_id=9833.

Hirsch, Joshua. *Afterimage: Film, Trauma, and the Holocaust.* Philadelphia: Temple University Press, 2003.

Hitler, Adolf. *Mein Kampf.* Translated by Ralph Manheim. New York: Houghton Mifflin Company, 1999.

Hobsbawm, E. J. *The Age of Extremes: A History of the World, 1914–1991.* New York: Pantheon Books, 1994.

Hutcheon, Linda. *A Poetics of Postmodernism: History, Theory, Fiction.* New York: Routledge, 1988.

Hynes, Samuel Lynn. *The Auden Generation: Literature and Politics in England in the 1930s*. New York: Viking Press, 1977.

Inge, William Ralph. *Christian Mysticism: The Bampton Lectures (1899)*. London: Methuen, 1912.

Isser, Edward. *Stages of Annihilation: Theatrical Representations of the Holocaust*. Madison: Fairleigh Dickinson University Press, 1997.

Jones, David Houston. "From Contumacy to Shame: Reading Beckett's Testimonies with Agamben." In *Beckett at 100*. Edited by Linda Ben-Zvi and Angela Moorjani. New York: Oxford University Press, 2008: 54–67.

Jones, Ernest. *Treatment of the Neuroses*. London: Ballière, Tindall and Cox, 1920.

Joyce, James. *Ulysses*. New York: Vintage Books, 1986.

Kalb, Jonathan. *Beckett in Performance*. Cambridge: Cambridge University Press, 1989.

Kearney, Richard. "Imagination, Testimony and Trust: A Dialogue with Paul Ricoeur." In *Questioning Ethics: Contemporary Debates and Philosophy*. Edited by Richard Kearney and Mark Dooley. London: Routledge, 1999.

Kennedy, Seán. "Ireland/Europe … Beckett/Beckett." In *Beckett and Ireland*. Edited by Seán Kennedy. Cambridge: Cambridge University Press, forthcoming.

Kenner, Hugh. *Flaubert, Joyce, and Beckett: The Stoic Comedians*. London: Dalkey Archive Press, 2005.

Kluger, Ruth. *Landscapes of Memory: A Holocaust Girlhood Remembered*. London: Bloomsbury, 2003.

Knowles, Dorothy. "Armand Gatti and the silence of the 1059 days of Auschwitz." In *Staging the Holocaust*. Edited by Claude Schumacher. Cambridge: Cambridge University Press, 1998.

Knowlson, James. *Damned to Fame: The Life of Samuel Beckett*. London: Bloomsbury, 1996.

———. "Foreword" to *The Beckett Country* by Eoin O'Brien. Dublin: Black Cat Press, 1986. x–xvi.

Knowlson, James and Elizabeth Knowlson, eds. *Beckett Remembering/Remembering Beckett*. New York: Arcade, 2007.

———, eds. *Beckett Remembering/Remembering Beckett*. London: Bloomsbury, 2006.

Knowlson, James, and John Pilling. *Frescoes of the Skull: The Later Prose and Drama of Samuel Beckett*. London: Calder, 1979.

Kristeva, Julia. *Black Sun: Depression and Melancholia*. New York: Columbia University Press, 1989.

Kubiak, Anthony. "Post Apocalypse Without Figures: The Trauma of Theater in Samuel Beckett." In *The World of Samuel Beckett*. Edited by Joseph H. Smith. Baltimore: Johns Hopkins University Press, 1990. 107–24.

Lacan, Jacques. *The Seminar of Jacques Lacan. Book II: The Ego in Freud's Theory and in Techniques of Psychoanalysis, 1954–1955*. Edited by Jacques-Alain Miller, translated by Sylvana Tomaselli. Cambridge: Cambridge University Press, 1988.

LaCapra, Dominick. *History in Transit: Experience, Identity, Critical Theory*. Edited by Moishe Postone and Eric Santner. Ithaca and London: Cornell University Press, 2004.

LaCapra, Dominick. "Holocaust Testimonies: Attending to the Victim's Voice." *Catastrophe and Memory: The Holocaust and the Twentieth Century.* Chicago: University of Chicago Press, 2003. 209–31.

———. *Writing History, Writing Trauma.* New York and Baltimore: Johns Hopkins University Press, 2001.

———. *Representing the Holocaust: History, Theory, Trauma.* Ithaca: Cornell University Press, 1996.

Lake, Carlton. *No Symbols Where None Intended.* Austin: Humanities Research Center, University of Texas, 1984.

Lakoff, George and Mark Johnson. *Metaphors we live by.* Chicago: University of Chicago Press, 1980.

Lamont, Rosette C. "Samuel Beckett's Wandering Jew." In *Reflections of the Holocaust in Art and Literature.* Edited by Randolph L. Braham. Boulder: Social Science Monographs, 1990. 35–53.

Langer, Lawrence. "Pursuit of Death in Holocaust Language." *Partisan Review* 3 (Summer 2001): 379–95.

Langer, Lawrence L. *Admitting the Holocaust: Collected Essays.* New York: Oxford University Press, 1975.

Laqueur, Walter, ed. *The Holocaust Encyclopedia.* Yale: Yale University Press, 2001.

Laughlin, Karen L. "'Dreaming of [...] Love': Beckett's Theatre and the Making of the (Post) Modern Subject." In "Endlessness in the Year 2000/Fin sans Fin en L'an 2000." *Samuel Beckett Today/Aujourd'hui* 11. Edited by Angela Moorjani and Carola Veit. Amsterdam: Rodopi, 2001. 202–9.

Leopardi, Giacomo. *Selected Prose and Poetry.* Translated by I. Origo and J. Heath-Stubbs. London: Oxford University Press, 1966.

Levi, Primo. *The Mirror Maker.* London: Abacus, 2002.

———. *Survival in Auschwitz: The Nazi Assault on Humanity.* New York: Simon & Schuster, 1996.

———. *The Drowned and the Saved.* Translated by Raymond Rosenthal. New York: Vintage, 1988.

Levinson, Marjorie. "What Is New Formalism?" *PMLA: Publications of the Modern Language Association of America* 122.2 (2007): 558–69.

Lewens, Tim. *Darwin.* New York: Routledge, 2007.

Lloyd, David. "Frames of *Referrance*: Samuel Beckett as an Irish Question." In *Beckett and Ireland.* Edited by Seán Kennedy. Cambridge: Cambridge University Press, forthcoming.

———. *Ireland After History.* Cork: Cork University Press, 2000.

Loustaunau-Lacau, Georges. *Chiens Maudits: Souvenirs d'un rescapé des bagnes hitlériens.* Paris: Editions du Réseau Alliance, 1945.

Lyotard, Jean-François. *The Differed: Phrases in Dispute.* Translated by Georges Van Den. Minneapolis: University of Minnesota Press, 1990.

Malkin, Jeanette. *Memory-Theater and Postmodern Drama.* Ann Arbor: University of Michigan Press, 1999.

McCormack, W.J. *From Burke to Beckett: Ascendancy, Tradition and Betrayal in Literary History.* Cork: Cork University Press, 1994.

———. "Seeing Darkly, Notes on T.W. Adorno and Samuel Beckett." In *Hermathena: A Trinity College Dublin Review* 141.2 (1986): 22–44.

McMullan, Anna. *Theatre on Trial: Samuel Beckett's Later Drama*. New York: Routledge, 1993.

McNaughton, James. "The Politics of Aftermath: Beckett, Modernism, and the Irish Free State." In *Beckett and Ireland*. Edited by Seán Kennedy. Cambridge: Cambridge University Press, forthcoming.

———. "Beckett, German Fascism, and History: The Futility of Protest." In "Historicising Beckett/Issues of Performance." *Samuel Beckett Today/Aujourd'hui* 15. Edited by Marius Buning, Matthijs Engleberts, Sjef Houppermans, and Dirk Van Hulle. Amsterdam: Rodopi, 2005. 101–16.

Mercier, Vivian. "The Uneventful Event." In *Critical Essays on Samuel Beckett: Critical Thought Series 4*. Edited by Lance St. John Butler. Aldershot: Scholar Press, 1993. 29–30.

Montague, John. "A Few Drinks and a Hymn: My Farewell to Samuel Beckett." *The New York Times*, Sunday Late Edition. 17 April 1994. http://www.samuel-beckett.net/beckett_hymn.html.

Moorjani, Angela. "Mourning, Schopenhauer, and Beckett's Art of Shadows." In *Beckett On and On*. Edited by Lois Oppenheim and Marius Buning. London: Farleigh Dickenson University Press, 1996. 83–101.

Muller-Doohm, Stefan. *Adorno: A Biography*. Translated by Rodney Livingstone. Cambridge: Polity Press, 2005.

Murphy, Peter (P.J.). *Reconstructing Beckett: Language for Being in Samuel Beckett's Fiction*. Toronto: University of Toronto Press, 1990.

Nietzsche, Friedrich Wilhelm. "Twilight of the Idols." In *The Portable Nietzsche*. Edited by Walter Kaufmann. New York: Penguin Books, 1982. 463–563.

Nixon, Mark. "Gospel und Verbot: Beckett und Nazi Germany." In *Das Raubauge in der Stadt: Beckett liest Hamburg*. Edited by Michaela Giesing, Gaby Hartel and Carola Veit. Göttingen: Wallstein Verlag, 2007. 79–88. Translated as "Ewangelia i zakaz: Beckett i nazistowskie Niemcy." *kwartalnik artystyczny* 1/2007 (53): 69–78.

———. "'Scraps of German': Samuel Beckett reading German Literature." In "Notes diverse holo." *Samuel Beckett Today/Aujourd'hui* 16. Edited by Matthijs Engelberts, Everett Frost, and Jane Maxwell. Amsterdam: Rodopi, 2006. 259–82.

———. "The *German Diaries* 1936/37: Beckett und die moderne deutsche Literatur." In *Der Unbekannte Beckett: Samuel Beckett und die deutsche Kultur*. Edited by Marion Dieckmann-Fries and Therese Seidel. Frankfurt a M.: Suhrkamp, 2005. 138–54.

———. "'What a Tourist I Must Have Been': The German Diaries of Samuel Beckett." Unpublished Ph.D. Thesis, University of Reading: 2005.

Nora, Pierre. *Realms of Memory: Rethinking the French Past*. Vol. 1. Columbia: Columbia University Press, 1996.

———. "Between Memory and History: *Les Lieux de Mémoire*." *Representations* 26 (Spring 1989): 7–25.

North, Michael. *The Political Aesthetic of Yeats, Eliot, and Pound*. Cambridge: Cambridge University Press, 1991.

O'Hara, J.D. "Beckett's Schopenhaurian Reading of Proust: The Will as Whirled in Re- presentation." In *Schopenhauer: New Essays in Honor of His 200th Birthday*. Edited by Eric van der Luft. Lewiston: The Edwin Mellen Press, 1988.

Olney, James. *Memory and Narrative: The Weaving of Life-Writing*. London: Chicago University Press, 1998.

Omer, Mordechai. *Avigdor Arikha: Drawings*. Tel Aviv: Museum of Art, 1998.

Perloff, Marjorie. " 'In Love with Hiding': Samuel Beckett's War." *Iowa Review* 35.2 (2005): 76–103.

———. *Wittgenstein's Ladder: Poetic Language and the Strangeness of the Ordinary*. Chicago: University of Chicago Press, 1996.

Peschanski, Denis. *La France Des Camps*. Paris: Gallimard, 2002.

Petrie, John. "The Secular Word 'HOLOCAUST': Scholarly Sacralization Twentieth Century Meanings." http://www.berkeleyinternet.com/holocaust/.

Phelan, Peggy. "Beckett and Avigdor Arikha." In *A Passion for Painting*. Edited by Fionnuala Croke. Dublin: Holberton, 2006.

Phillips, Adam. "Close-Ups." *History Workshop Journal* 57 (2004): 142–49.

Pick, Daniel. *Faces of Degeneration: A European Disorder, c. 1848–1918*. Cambridge: Cambridge University Press, 1989.

Pilling, John. *A Samuel Beckett Chronology*. London: Palgrave, 2006.

———. *Beckett's Dream Notebook*. Reading: Beckett International Foundation, 1999.

———. "Beckett's *Proust*." Reprinted in *The Beckett Studies Reader*. Edited by S.E. Gontarski. Gainesville: University Press of Florida, 1993. 1–28.

Pountney, Rosemary. *Theatre of Shadows: Samuel Beckett's Drama 1956–76*. Gerrards Cross/Totowa: Colin Smythe/Barnes and Noble Books, 1988.

Rabinovitz, Rubin. *The Development of Samuel Beckett's Fiction*. Chicago: University of Illinois Press, 1984.

Razac, Olivier. *Barbed Wire: A Political History*. New York: New Press: Distributed by W.W. Norton, 2002.

Restivo, Giuseppina. "*Melencolias* and Scientific Ironies in *Endgame*: Beckett, Walther, Dürer, Musil." In "Endlessness in the Year 2000/Fin sans Fin en L'an 2000." *Samuel Beckett Today/Aujourd'hui* 11. Edited by Angela Moorjani and Carola Veit . Amsterdam: Rodopi, 2002. 103–13.

———. "The Genesis of Beckett's *Endgame* Traced in a 1950 Holograph." In "Intertexts in Beckett's Work et/ou Intertextes de l'oeuvre de Beckett." *Samuel Beckett Today/Aujourd'hui* 3. Edited by Marius Buning and Sjef Houppermans. Amsterdam: Rodopi, 1994. 85–96.

Richter, Gerhard. "Acts of Memory and Mourning: Derrida and the Fictions of Anteriority." In *Mapping Memory*. Edited by Susannah Radstone and Bill Schwarz. New York: Fordham University Press, forthcoming.

Ricks, Christopher, ed. *The Oxford Book of English Verse*. Oxford: Oxford University Press, 1999.

Ricoeur, Paul. *History, Memory, Forgetting*. Chicago: Chicago University Press, 2006.

————. "Memory and Forgetting." In *Questioning Ethics: Contemporary Debates in Philosophy*. Edited by Richard Kearney and Mark Dooley. London: Routledge, 1999. 5–11.

Rose, Jacqueline. "Virginia Woolf and the Death of Modernism." In *On Not Being Able to Sleep: Psychoanalysis and the Modern World*. London: Vintage, 2004. 72–88.

Rosen, Steven J. *Samuel Beckett and the Pessimistic Tradition*. New Brunswick: Rutgers University Press, 1976.

Roudané, Matthew. *Dramatic Essentials*. Boston and New York: Houghton Mifflin, 2009.

Ryan, Kiernan. *New Historicism and Cultural Materialism: A Reader*. London: Arnold, 1996.

Said, Edward. *Reflections on Exile and Other Essays*. Cambridge, MA: Harvard University Press, 2000.

Sartre, Jean Paul. *What Is Literature*. London: Methuen, 1967.

Scarry, Elaine. *The Body in Pain: The Making and Unmaking of the World*. New York and Oxford: Oxford University Press, 1985.

Schopenhauer, Arthur. *On Vision and Colors: An Essay*. Edited by David E. Cartwright and translated by E.F.J. Payne. Oxford: Berg, 1994.

————. *The World as Will and Idea*. Vol. 1, Book III. Translated by Richard Burdon Haldane and John Kemp. London: Kegan Paul, Trench, Trübner, 1907.

————. *The World as Will and Representation*. Translated by E.F.J. Payne. New York: Dover Publications, 1969.

Seibers, Tobin. "Kant and the Politics of Beauty." *Philosophy and Literature* 22.1 (1998): 31–50.

Shelley, Mary Wollstonecraft. *Frankenstein*. London: Penguin Classics, 1992.

Sherry, Vincent B. *The Great War and the Language of Modernism*. New York: Oxford University Press, 2003.

Shildo, Anna. "Tragic Elements in Samuel Beckett's Dramatic Theory and Practice." Unpublished Ph.D. Thesis, Tel Aviv University, 1993.

Simpson, David. *9/11: The Culture of Commemoration*. Chicago: University of Chicago Press, 2006.

Spivak, Gayatri Chakravorty. "The Staging of Time in *Heremakhonon*." In Teresa Heffernan and Jill Didur, eds. *Cultural Studies* 17.1 (2003): 85–97.

Stanford, W.B. *Memoirs*. Dublin: Hinds, 2001.

Steiner, George. "Silence and the Poet." In *Language and Silence and Other Essays*. New Haven: Yale University Press, 1988. 36–54.

————. *The Death of Tragedy*. London: Faber & Faber, 1978.

Stewart, Paul. "The Need for Beckett." In *Other Becketts*. Edited by Daniela Caselli, Steven Connor, and Laura Salisbury. Tallahassee: Journal of Beckett Studies Books, 2002. 17–29.

Stieve, Friedrich. "What the World Rejected: Hitler's Peace Offers." *Washington Journal*. 1940.

————. *Abriss der Deutschen Geschicte von 1792–1935*. Leipzig: Kohlhammer, 1937.

———. *Geschichte Des Deutschen Volkes*. München and Berlin: Verlag von R. Oldenbourg, 1934.

Strier, Richard. "How Formalism Became a Dirty Word, and Why We Can't Do Without It." Afterword to *Renaissance Literature and Its Formal Engagements*. Edited by Mark David Rasmussen. New York: Palgrave, 2002.

Uhlmann, Anthony. *Beckett and Poststructuralism*. Cambridge: Cambridge University Press, 1999.

Van Hulle, Dirk. "'Nichtsnichtsundnichts': Beckett's and Joyce's Transtextual Undoings." In *Beckett, Joyce and the Art of the Negative. European Joyce Studies* 16. Edited by Colleen Jaurretche. Amsterdam: Rodopi, 2005. 49–62.

Van Hulle, Dirk and Mark Nixon. "Holo and unholo." In "All Sturm and No Drang." *Samuel Beckett Today/Aujourd'hui* 18. Edited by Dirk Van Hulle and Mark Nixon. Amsterdam: Rodopi, 2007. 313–22.

Vinen, Richard. *The Unfree French: Life Under the Occupation*. London: Penguin, 2007.

Weiler, Gershon. *Mauthner's Critique of Language*. Cambridge: Cambridge University Press, 1970.

Weiss, Katherine. "Bits and Pieces: The Fragmented Body in *Not I* and *That Time*." In *Other Becketts*. Edited by Daniela Caselli, Steven Connor, and Laura Salisbury. Tallahassee: Journal of Beckett Studies Books, 2002. 187–95.

Weschler, Judith. "Illustrating Samuel Beckett." In *Art Journal* 10.23 (Winter 1993, http://findarticles.com/p/articles/mi_m0425/is_n4_v52/ai_14970137/pg_4.

Wills, Clair. *That Neutral Island: A Cultural History of Ireland during the Second World War*. London: Faber & Faber, 2007.

Wilson, Ann. "'Her Lips Moving': The Castrated Voice of *Not I*." In *Women in Beckett: Performance and Critical Perspectives*. Edited by Linda Ben-Zvi. Urbana: University of Illinois Press, 1990. 190–200.

Wood, Michael. "Vestiges of Ireland in Beckett's Late Fiction." In *Beckett and Ireland*. Edited by Seán Kennedy. Cambridge: Cambridge University Press, forthcoming.

Žižek, Slavoj. *On Belief.* London: Routledge, 2001.

———. *Looking Awry*. Cambridge and London: MIT Press, 1992.

———. *The Sublime Object of Ideology*. London; New York: Verso, 1989.

Contributors

Jackie Blackman is an Irish Research Council of Humanities and Social Sciences scholar at the Samuel Beckett Centre, Trinity College, Dublin. She has lectured on Beckett at Trinity College and the Centre Culturel Irlandais in Paris. In 2006 she was dramaturge for a centenary production of *Endgame* at the Samuel Beckett Theatre in Dublin. Her article "Beckett Judaizing Beckett: 'A Jew from Greenland' in Paris" has been published in *Samuel Beckett Today/Aujourd'hui* (2007).

Jonathan Boulter is an Associate Professor of English at the University of Western Ontario. He is the author of *Interpreting Narrative in the Novels of Samuel Beckett* (University Press of Florida, 2001), *Samuel Beckett: A Guide for the Perplexed* (London: Continuum Press, 2008), and several articles on Beckett. His essay "Does Mourning Require a Subject? Samuel Beckett's *Texts for Nothing*" (*Modern Fiction Studies*) won the Margaret Church MFS Memorial Prize 2004.

Matthew Feldman is a lecturer in Twentieth Century History at the University of Northampton, and the editor of the Routledge political science journal, *Totalitarian Movements and Political Religions*. He has also published widely on Samuel Beckett, including essays in *Journal of Beckett Studies, Samuel Beckett Today/Aujourd'hui* as well as his 2006 monograph, *Beckett's Books: A Cultural History of Samuel Beckett's 'Interwar Notes'* (Continuum Books). He runs the ongoing seminar series in Oxford, England, *Samuel Beckett: Debts and Legacies*. He has published *Beckett's Literary Legacies* and *Beckett's International Reception* (both with Mark Nixon, 2007/2009), *Beckett and Phenomenology* (with Ulrika Maude, 2008) and *Beckett and Death* (2008).

Alysia E. Garrison is a Ph.D. candidate at the University of California, Davis. She has recently published an article, "'Disdaining Bounds of Place and Time,' Staining Language with Furze and Burvine: John Clare's

Nomadic Poetics," which appeared in *Literature Compass* (2006). Her work on mourning and trauma has landed her several honors and awards including the Robert Stoller Foundation Student Fellow (2005).

Dirk Van Hulle is an Associate Professor of Literature in English at the University of Antwerp, Belgium. He is executive editor of the series of genetic editions of Samuel Beckett's bilingual works and is preparing an electronic genetic edition of Beckett's last works (Brepols Publishers, 2006). He is author of *Textual Awareness: A Genetic Study of Late Manuscripts by Joyce, Proust and Mann* (University of Michigan Press, 2004) and recently edited the volume of essays *Beckett the European* (*Journal of Beckett Studies*, 2005). He maintains the Beckett Endpage (www.ua.ac.be/Beckett) and is currently working with Mark Nixon and Vincent Neyt on a digital manuscript edition of four works by Samuel Beckett.

Seán Kennedy is an Associate Professor of English at St Mary's University, Halifax, Nova Scotia. He is editor of *Beckett and Ireland* (Cambridge University Press, forthcoming).

James McNaughton currently teaches at the University of Alabama, both as a Senior Fellow in the Blount Undergraduate Initiative and, beginning in 2007, as an Assistant Professor in the Department of English. He has one article published on Samuel Beckett and another forthcoming. McNaughton's interests more broadly include European Modernisms, twentieth century Irish literature, and the relationship among aesthetics, politics, and history. He is currently writing a book on Beckett's early politics.

Mark Nixon is the Lecturer in English at the University of Reading, where he is also the Co-Director of the Beckett International Foundation. He has published more than twenty essays on Beckett's works, and has recently edited, with Matthew Feldman, *The International Reception of Samuel Beckett* (Continuum, 2009). He is an editor of *Samuel Beckett Today/Aujourd'hui*, reviews editor of the *Journal of Beckett Studies*, and the Co-Director of the Beckett Digital Manuscript Project. He is currently working on *Beckett's Library* with Dirk Van Hulle and is editing a book on *Beckett and Publishing* (British Library, 2010). He is also preparing, for Faber and Faber, an edition of *Beckett's Shorter Fiction 1950–1981*, and also for Faber, a critical edition of the unpublished short story "Echo's Bones."

Robert Reginio, a recent Ph.D. graduate from the University of Massachusetts-Amherst, currently teaches drama and literature at his alma mater. In 2006, he both received the Harry Ransom Humanities Research Center Dissertation Fellowship and placed second for the International Federation of Theatre Research: New Scholar's Prize.

Katherine Weiss is an Assistant Professor of English at East Tennessee State University. She has published essays on Beckett's short prose fiction and drama including "De/composing the Machine in Samuel Beckett's *The Lost Ones* and *Ping*" (*Stirrings Still*, 2004) and "Bits and Pieces: The Fragmented Body in *Not I* and *That Time*" (*Journal of Beckett Studies*, 2001), and is currently working on a book length study entitled *Breaking Down: Technology in the Works of Samuel Beckett.*

Index

Lightning Source UK Ltd.
Milton Keynes UK
UKOW04n0741090316

269884UK00013B/238/P